CHEMISTRY RESEARCH AND APPLICATIONS

MOLECULAR MECHANICS AND MODELING

CHEMISTRY RESEARCH AND APPLICATIONS

Additional books in this series can be found on Nova's website under the Series tab.

Additional e-books in this series can be found on Nova's website under the e-book tab.

CHEMISTRY RESEARCH AND APPLICATIONS

MOLECULAR MECHANICS AND MODELING

PRISCILLA WATKINS
EDITOR

nova
publishers
New York

NOTICE TO THE READER

The Publisher has taken reasonable care in the preparation of this book, but makes no expressed or implied warranty of any kind and assumes no responsibility for any errors or omissions. No liability is assumed for incidental or consequential damages in connection with or arising out of information contained in this book. The Publisher shall not be liable for any special, consequential, or exemplary damages resulting, in whole or in part, from the readers' use of, or reliance upon, this material. Any parts of this book based on government reports are so indicated and copyright is claimed for those parts to the extent applicable to compilations of such works.

Independent verification should be sought for any data, advice or recommendations contained in this book. In addition, no responsibility is assumed by the publisher for any injury and/or damage to persons or property arising from any methods, products, instructions, ideas or otherwise contained in this publication.

This publication is designed to provide accurate and authoritative information with regard to the subject matter covered herein. It is sold with the clear understanding that the Publisher is not engaged in rendering legal or any other professional services. If legal or any other expert assistance is required, the services of a competent person should be sought. FROM A DECLARATION OF PARTICIPANTS JOINTLY ADOPTED BY A COMMITTEE OF THE AMERICAN BAR ASSOCIATION AND A COMMITTEE OF PUBLISHERS.

Additional color graphics may be available in the e-book version of this book.

Library of Congress Cataloging-in-Publication Data

ISBN: 978-1-63483-388-2

Published by Nova Science Publishers, Inc. † New York

CONTENTS

PREFACE

This book focuses on the latest developments in molecular mechanics and modeling research. Included in this book are chapters reviewing the progress of the bridging cell method (BCM) in its formulation and its nanostructure and material modeling applications; an analysis of the reaction selectivity of alanine racemase mutants; and a step-wise process for inferring the stereo structure of proteins constructed using related stereo structures. Lastly, molecular mechanics based models have proven to provide a representation of average carbon porosity and enable computational estimates of molecule-surface interaction energies. In the last chapter, the authors focus on graphene layer pore models for molecule-surface binding energies.

Chapter 1 - In this chapter, the authors review the progress of the bridging cell method (BCM) in its formulation and application to nanostructure and material modeling applications. The issues of nonlocal/local energy mismatch at atomistic/continuum boundaries along with the issues of temperature effects and timescale concerns are detailed, as well as how the BCM addresses these issues. Previous applications of the BCM to the simulation of nanovoids and cracks, nanoindentation, and deformation in polymeric materials are reviewed. In each case, the multiscale models have agreed well with full molecular dynamics (MD) simulations and experimental work, while showing a significant reduction in simulation time. In this respect, the BCM has proven to be a viable alternative to time-consuming MD methods, which are not applicable to many engineering applications. The chapter will also overview the future direction of the BCM to address new applications as well as outstanding issues that still exist within the concurrent multiscale modeling field.

Chapter 2 - Experimental results regarding the contribution of alkyl chains towards hydrogen-bonding capability of hydroxyl groups were analyzed using different versions of the MM2 program. The contribution of hydrogen-bonding energy values calculated using the MM2 program of the current CAChe programs differed significantly from those of previous ones, and the dihedral angle energy value was found to affect the contribution of alkyl chain length. The current report elaborates a step-wise process for inferring the stereo structure of proteins constructed using related stereo structures in spite of a less than 35% amino acid sequence similarity. A quantitative analysis of enzyme reactivity was performed on proteins such as a mutant of R-amino acid oxidase M215R and aromatic alcohol $NADP^+$ oxidoreductase ELIS-2.

Chapter 3 - Downloaded structure of alanine racemase was optimized using a molecular mechanics calculation and used as an initial enzyme, then arginine-219 was replaced with other amino acids to construct the mutants. The atomic distances, bond angles, and apc of the mutants were correlated with enzyme reactivity and used to study the enzyme reaction mechanism. The alanine carboxyl group contacted lysine-39 and arginine-136. The alanine amino group contacted the oxygen molecules of pyridoxal 5'-phosphate aldehyde and phosphate, and aspartic acid-171. These amino acid residues contacted alanine with Coulombic force and may contribute to the enolation, rather than directly affecting the α-hydrogen exchange. The alanine α-hydrogen seldom contacted tyrosine-265'. The enzyme reactivity was quantitatively related to the bond angle of the substituted alanine. Although the apc can be used for quantitative analysis, the slope was small to predict the selective reactivity of the mutants. Which tyrosine is directly involved in the reaction seems to depend on the selected enzyme and mutant. A limited molecular dynamic calculation supported the results obtained using the molecular mechanics calculation.

Chapter 4 - Gas-solid chromatography experiments were used to determine the second gas-solid virial coefficient (B_{2s}) values for 10 adsorbate molecules interacting with the solid adsorbent Carbosieve S-III (Supelco). The adsorbates used included: methane, ethane, chloromethane, chlorodifluoromethane, 1,1-difluoroethane, propane, hexafluoroethane, dichlorodifluoromethane, 1-chloro-1,1-difluoroethane, and butane. Carbosieve S-III is a porous carbon powder with a fairly uniform slit width pore size of 0.55 nm and a Nitrogen BET surface area of 995 m^2 /g. The B_{2s} data were collected over a range of temperature and used to construct van't Hoff plots of $\ln(B_{2s})$ versus 1/T for each molecule. The slopes of these plots gave experimental molecule-surface binding energies or adsorption energies (E*).

Computational experiments using molecular mechanics with MM2 parameters were used to generate surface models and calculate the binding energy (Ecal*) and a simple modification of the same ($Ecal^*_{modified}$) based on the fraction of nonhydrogen sp^3 carbon atoms in the adsorbate molecule. Ten different models were constructed: plate, parallel plate (D = 0.94 nm, 0.89 nm, 0.84 nm, 0.79 nm, 0.74 nm), and double parallel plate (D = 1.04 nm, 0.99 nm, 0.89 nm, 0.84 nm) where D is the internuclei separation measured from the innermost carbon atoms in the graphene layer of one plate to the innermost carbon atoms of the graphene layer of the other plate. Each plate in the model was constructed of three parallel graphene layers and each graphene layer consisted of 127 hexagonal benzene ring structures. Thus the plate model has one wall with one set of three layers, the parallel plate model has two walls with two sets of three layers, and the double parallel plate model has four walls with four sets of three layers. This last model forms a box-like structure that is open at the top and bottom. Of the 10 models, the double parallel plate model with internuclei parallel wall separation of D = 0.89 nm (interior of 0.55 nm across) correlated best with the experimental molecule-surface binding energies. Experimental values versus these calculated and modified values of molecule-surface binding energies gave a slope of 1.025 with a correlation of $R^2 = 0.978$. Molecular mechanics based models can provide a representation of average carbon porosity and enable computational estimates of molecule-surface interaction energies.

In: Molecular Mechanics and Modeling
Editor: Priscilla Watkins
ISBN: 978-1-63483-388-2

Chapter 1

FINITE ELEMENT BASED MULTISCALE MODELING USING THE BRIDGING CELL METHOD

Vincent Iacobellis and Kamran Behdinan*
Department of Mechanical & Industrial Engineering,
University of Toronto, Toronto, ON, Canada

ABSTRACT

In this chapter we review the progress of the bridging cell method (BCM) in its formulation and application to nanostructure and material modeling applications. The issues of nonlocal/local energy mismatch at atomistic/continuum boundaries along with the issues of temperature effects and timescale concerns are detailed as well as how the BCM addresses these issues. Previous applications of the BCM to the simulation of nanovoids and cracks, nanoindentation, and deformation in polymeric materials are reviewed. In each case the multiscale models have agreed well with full molecular dynamics (MD) simulations and experimental work while showing a significant reduction in simulation time. In this respect, the BCM has proven to be a viable alternative to time consuming MD methods which are not applicable to many engineering applications. The chapter will also overview the future

* Contacting author email: behdinan@mie.utoronto.ca.

direction of the BCM to address new applications as well as outstanding issues that still exist within the concurrent multiscale modeling field.

1. INTRODUCTION

The recent growth in fields such as nanotechnology and advanced nanomaterials research has created the need to accurately model and analyze processes which occur on multiple scales. It has been shown that small scale effects at the atomistic level are the basis for problems such as surface effects in NEMS/MEMs devices, fabrication of nanomaterials, and dislocation/ fracture growth in composite structures. In such cases simulation must be performed on the atomic level; however this can lead to computational requirements that are not feasible. In order to model the large scale response to the small scale behavior a class of multidisciplinary modeling techniques have been developed termed multiscale modeling. As shown in Figure 1, multiscale methods can be divided into two main groups: hierarchal and concurrent. Hierarchal approaches [1-3] pass the atomistic properties of the solid to the continuum where the governing equations are obtained through a homogenization procedure that involves the transfer of parameters such as deformation and stress fields. Concurrent multiscale methods [4] 'embed' the atomistic model directly into the continuum by dividing the system into separate atomistic and continuum domains coupled by a transitional domain.

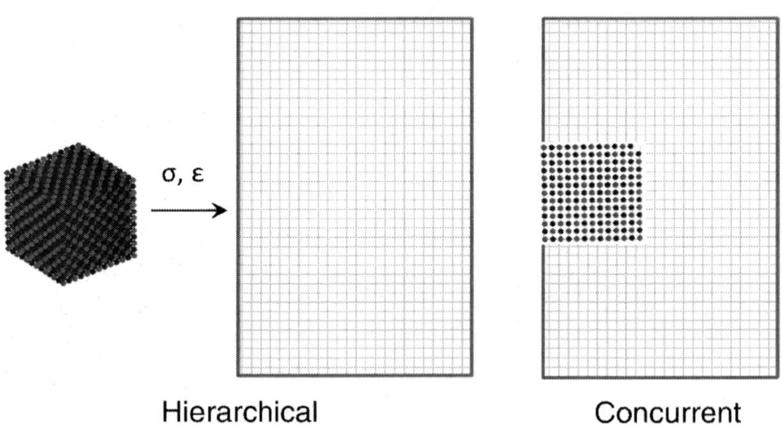

Figure 1. Hierarchical and concurrent multiscale approaches.

An alternative concurrent multiscale coupling approach has been developed, termed the bridging cell method (BCM) [5], whose basic framework has been applied to a number of applications in both crystalline and amorphous material structures. The BCM is a multiscale technique which formulates atomistic, bridging, and continuum domains all in a common finite element framework. The BCM couples atomistic and continuum formulations through a bridging domain that consists of a series of cells whose underlying atomic structure is interpolated to the nodes of each cell through mapping functions. Through each domain, a temperature dependent potential is incorporated to account for temperature effects, thus avoiding the small timesteps associated with molecular dynamics (MD).

This chapter will review the progress of the BCM in its formulation and application to nanostructure and material modeling applications. Results will be presented with respect to the BCM's previous application to the simulation of nanovoids and cracks [6, 7], nanoindentation [8], and deformation in polymeric materials. The first section will provide a brief review of concurrent multiscale modeling approaches as well as some of the existing challenges. The next section will provide an overview of the BCM along with the extent of its application space, followed by an overview of the future direction of the BCM to address new applications and areas of future and active research.

2. Review of Concurrent Multiscale Methods

Concurrent multiscale methods involve decomposing the system into three domains: an atomistic domain, a continuum domain, and a bridging/handshake domain which acts to couple the atomistic and continuum formulations. There are typically two methods for coupling continuum and atomistic domains. One such approach is to resolve the continuum finite element mesh down to the scale of the atomic spacing at the interface between the atomistic and continuum domains. The transfer of information across the interface is performed by constraining the finite element nodes to move with the atoms at the interface. In this sense, the coupling domain creates a strict boundary condition between the finite element nodes and the atoms.

One such technique is the quasicontinuum (QC) method, developed by Tadmor et al. [9], which employs the use of representative atoms or 'repatoms'. Repatoms are used to represent a collection of atoms undergoing similar deformation as opposed to tracking the deformation of each atom separately. In regions where nonlinear behavior dominates, the locations of the repatoms correspond to the atomic lattice sites and incorporate the nonlocal energy of the atomistic formulation. In the regions where a continuum formulation is sufficient, each repatom represents the deformation of a collection of atoms. Thus, by only considering a select number of repatoms in the simulation the total number of degrees of freedom is reduced. A number of variations and advancements have been made with regard to the QC method such as the cluster-based formulation [10, 11], mesh adaptability [12], application to three dimensional deformation problems [13, 14], and finite temperature extensions [15-19]. Similar approaches to the QC method include the coupling of length scales (CLS) method [20], finite element atomistics method (FEAt) [21], and the coupled atomistic discrete dislocation (CADD) method [22], which allows for the passing of dislocations from the atomistic domain to a discrete dislocation dynamics informed continuum domain.

The other approach to concurrent multiscale coupling is to incorporate an overlapping/handshake domain where the finite element mesh is not resolved down to the atomic spacing at the interface, but rather both formulations overlap each other and are coupled using an averaging approach. In this sense, the displacements of the atoms and finite element nodes in this overlapping domain can be considered as coupled in a weak sense. One such approach is the bridging domain method (BDM) developed by Xiao and Belytschko [23]. In the BDM, the potential energy of the handshake domain consists of weighted contributions from both continuum and atomistic descriptions based on a partition of unity scheme. Compatibility in the overlapping, or handshaking domain, is enforced by Lagrange multipliers based on the approach used in the Arlequin method [24, 25]. The handshake domain consists of a linear scaling of the energies where the energy from the continuum is dominant near the handshake/continuum interface and the energy from the atomistic domain is dominant at the other handshake/atomistic interface. As a result of the energy scaling, the position of atoms does not have to be coincident with nodes in the continuum mesh, avoiding complex meshing procedures associated atomic scale mesh resolution.

The BDM has been applied extensively in the analysis of the mechanical behavior of carbon nanotubes and graphene sheets [26]. Recently the method has been applied to modeling crack surfaces and slip planes by using the BDM scheme to couple an MD simulation with a continuum region described by the Extended Finite Element Method (XFEM) [27]. The method has also been used in coupling quantum mechanics, molecular dynamics, and continuum mechanics simulations by Khare et al. [28]. Xu et al. [29] developed the relaxed BDM for composite lattices. A similar method, the AtC method [30-

33], takes the same Arlequin based coupling approach, however employs a force based formulation as opposed to the energy approach used in the BDM.

Another handshake approach is that of the bridging scale method (BSM) developed by Liu et al. [34, 35]. In the BSM, the continuum finite element domain exists everywhere in the system. In the region where both finite element nodes and atoms overlap, the finite element nodal displacements are determined by the least squares fit of the underlying atomistic displacements. The BSM uses an analytically and numerically derived impedance force at the atomistic/continuum boundary to eliminate spurious phonon reflections at the interface and allow them to propagate into the overlapping continuum domain. The BSM has been applied to various applications including the quasistatic analysis of carbon nanotubes [36] as well as 2D and 3D crack propagation [37].

The above description of some of the existing multiscale methods is not extensive. Many others exist including the perfectly matched multiscale method [38], heterogeneous multiscale method [39, 40], embedded statistical coupling method [41], and coarse grained molecular dynamics (CGMD) [42]. For more review on existing concurrent multiscale methods one can refer to [43, 44] as well as [45] and [4] for a comparison of existing techniques.

3. Existing Issues

Although the above coupling strategies have been applied successfully to a number of different applications, there are still issues that exist in the multiscale modeling field. One such issue is the presence of unnatural forces at the atomistic/continuum domain interface due to the mismatch in the energy formulation between the two domains. The energy in the continuum domain can be modeled using a local description where the energy at a particular location is dependent upon the parameters at the location alone. This differs from the atomistic domain where the energy is calculated using a nonlocal energy formulation since the energy of an atom is dependent upon not only the properties at that atom's location, but also the properties of its neighboring atoms. Since atoms close the atomistic/continuum interface do not have full set of neighboring atoms to interact with, the result is the formation of spurious forces, also known as ghost forces, at the interface. Numerous techniques have been developed to mitigate these forces including the use of dead loads applied to atoms at the interface [12, 46], force based formulations [21, 22, 32], projection methods [42, 47], and the use of quasi-nonlocal atoms [48, 49].

Another issue associated with atomistic/continuum domain interface incompatibility occurs in simulations involving MD. When using MD, displacement waves will form in the atomistic domain that cannot be dissipated into the continuum domain and in turn are reflected back into the atomistic domain. MD is also used for many applications in order to run the simulation at finite temperature since MD explicitly tracks the velocities of the atoms. However, to do so requires small timesteps (on the order of femto- and pico-seconds), thus making it not applicable to a number of practical engineering problems. To address this issue, possible solutions have been proposed including multiple timestepping [50] as well as integrating the temperature into the interatomic energy function either directly [51, 52] or through consideration of the vibrational modes of the atomistic system [53, 54].

4. Bridging Cell Method

4.1. Energy Description in the BCM

The BCM multiscale approach couples atomistic-to-continuum scales through an overlapping bridging domain as shown in Figure 2. The bridging domain consists of cells whose constitutive formulation is derived from the underlying atoms contained within each cell [5]. The atom displacements within each cell are mapped to the deformation of the cell based on a set of mapping, or basis, functions. In this sense there is strict coupling between the atomistic and continuum domains without the need to resolve the finite element mesh down to the atomic spacing. The atomistic domain is reformulated into a finite element discretization [55] such that each domain can be solved using a finite element framework. This reformulation allows for established finite element solvers to be used in the solution process, facilitating both implementation and speed-up of the multiscale method.

The overall energy of the system, E_{tot}, is composed of nonlocal and local energy contributions [5],

$$E_{tot} = E_C^L + E_B^L + E_B^{NL} + E_A^{NL} , \tag{1}$$

where subscripts C, B, and A correspond to the continuum, bridging, and atomistic domain, respectively, and superscripts NL and L correspond to nonlocal and local energy formulations, respectively. Thus, the atomistic domain consists of a nonlocal energy contribution, the continuum domain consists of a local energy contribution, and in the bridging domain the energy is composed of both local and nonlocal energy contributions [5]. Therefore, the energy in the bridging domain can be decomposed into,

$$E_B^L = \sum_{c=1}^{n_{cell}} \sum_{q=1}^{n_q} w_q^{(c)} \psi(\mathbf{F}_c) V_c \quad , \qquad E_B^{NL} = \sum_{c=1}^{n_{cell}} \sum_{i=1}^{n_{aB}} U_i \tag{2}$$

where ψ is the strain energy density as a function of the deformation gradient of the cell $\mathbf{F}_c$, V_c is the volume of cell c, n_{cell} is the number of cells, U_i is the site energy for atom i, and n_{aB} and n_{nB} are the number of atoms and number of nodes located inside the bridging domain, respectively. The weighting, $w_q^{(c)}$, assigned to the quadrature points of each cell is derived from the ratio of the total interatomic energy contained within the cell to the total local energy of the entire cell. For pairwise potentials this can be represented by [7],

$$w^{(c)} = 1 - \frac{\sum_{i \neq j} R_{ij}^c U(\mathbf{r}_{ij})}{\psi(\mathbf{F}_c) \cdot V_c}, \tag{3}$$

where $R_{ij}^{\ c}$ is the ratio of the bond between atoms i and j located within cell c. The numerator in the second term of Eq. (3) represents the amount of strain energy within the cell, while the denominator represents the total strain energy located in the cell if each atom in the cell had a complete set of neighboring atoms to interact with. For multibody potentials, such as the embedded atom method (EAM), Eq. (3) provides an approximation to the correct weighting of the energy in the bridging domain [6]. If the system involves additional interactions, such as in the case of polymeric materials, Eq. (3) can be rewritten as,

$$w^{(c)} = 1 - \frac{1/2\mathbf{u}_c^T \mathbf{K}_c^{NL} \mathbf{u}_c}{\psi_c(\mathbf{F}_c) \cdot V_c}. \tag{4}$$

In Eq. (4) $\mathbf{u}_c$ are prescribed displacements for cell c, while K_c^{NL} is the nonlocal stiffness matrix of cell c as defined in Sec. 4.2. By straining each cell and comparing the strain energy based on the nonlocal stiffness (the numerator in Eq. (4)) to the strain energy of the cell based on a local energy formulation (denominator in Eq. (4)), the weighting between atomistic and continuum energies can be calculated such that the internal forces acting on the cell nodes are zero under conditions of uniform strain. In other words the deformation between the continuum and atomistic domains will be consistent.

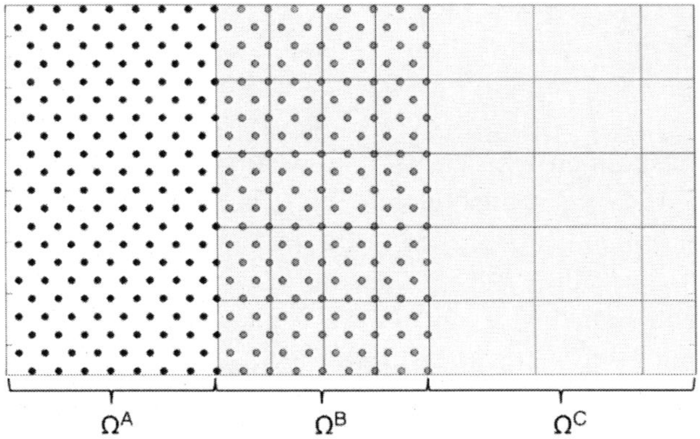

Figure 2. BCM domains: atomistic domain ($\boldsymbol{\Omega}^A$), bridging domain ($\boldsymbol{\Omega}^B$), and continuum domain ($\boldsymbol{\Omega}^C$).

4.2. Finite Element Framework of the BCM

Presented here is a short overview of the implementation of the BCM. For a complete derivation of the BCM in a finite element framework one can refer to [5]. The finite element method provides an efficient means of finding the state of minimum energy for a system. The minimum energy corresponds to setting the derivative of the nonlocal energy with respect to the configuration, $\mathbf{x}$, to zero,

$$\frac{\partial E_{tot}}{\partial \mathbf{x}} = 0 \,. \tag{5}$$

Linearizing Eq. (5) through the Taylor expansion of E_{tot} around the initial configuration, $\mathbf{x}_0$, the following is obtained [55],

$$E_{tot} \approx E_{tot}(\mathbf{x}_0) + \left.\frac{\partial E_{tot}}{\partial \mathbf{x}}\right|_{\mathbf{x}=\mathbf{x}_0} \cdot (\mathbf{x}-\mathbf{x}_0) + \frac{1}{2}(\mathbf{x}-\mathbf{x}_0)^T \cdot \left.\frac{\partial^2 E_{tot}}{\partial \mathbf{x}\partial \mathbf{x}}\right|_{\mathbf{x}=\mathbf{x}_0} \cdot (\mathbf{x}-\mathbf{x}_0) \cdot \tag{6}$$

Substituting Eq. (6) into Eq. (5),

$$\left.\frac{\partial^2 E_{tot}}{\partial \mathbf{x}\partial \mathbf{x}}\right|_{\mathbf{x}=\mathbf{x}_0} \cdot \Delta\mathbf{x} = -\left.\frac{\partial E_{tot}}{\partial \mathbf{x}}\right|_{\mathbf{x}=\mathbf{x}_0}, \tag{7}$$

$$\mathbf{K}\Delta\mathbf{x} = \mathbf{F} - \mathbf{f}$$

where $\Delta\mathbf{x} = \mathbf{x}\text{-}\mathbf{x}_0$ are the atomic and nodal displacements and **F** and **f** are the additional external forces and internal forces respectively. The stiffness matrix in Eq. (7) is made up of local and nonlocal stiffness matrices associated with the atomic and element stiffness respectively. The BCM couples the two formulations through the bridging domain where the displacements of the atoms in the bridging domain are mapped to the displacements of the bridging cells. This mapping is done through mapping functions, **M**, which thus far have been equivalent to linear shape functions used in the finite element method and dependent on the geometry and number of nodes of the bridging cell [5]. If the resulting mapping produces high energy configurations, atoms in the bridging domain are allowed to relax freely once the local continuum and bridging domain residual is below a set tolerance. Once the energy has been minimized sufficiently, the atoms in the bridging domain are remapped to the bridging elements and the simulation continues [6]. Incorporating the mapping function into Eq. (7), the coupled finite element equation is represented by,

$$\left[\mathbf{K}_{CB}^{L} + \mathbf{M}^T\mathbf{K}_{AB}^{NL}\mathbf{M}\right]\Delta\mathbf{x} = \left\{\mathbf{F}_{CB}^{L} + \mathbf{M}^T\mathbf{F}_{AB}^{NL}\right\} - \left\{\mathbf{f}_{CB}^{L} + \mathbf{M}^T\mathbf{f}_{AB}^{NL}\right\}, \tag{8}$$

where the $\mathbf{K}_{CB}{}^{L}$ is the stiffness matrix corresponding the local energy formulation of the continuum and bridging domains, $\mathbf{K}_{AB}{}^{NL}$ is the stiffness matrix corresponding to the nonlocal energy formulation of the atomistic and bridging domains, $\mathbf{F}_{CB}{}^{L}$ and $\mathbf{f}_{CB}{}^{\mathrm{L}}$ are applied and internal forces acting on the continuum and bridging domain nodes, respectively, and $\mathbf{F}_{AB}{}^{NL}$ and $\mathbf{f}_{AB}{}^{NL}$ are applied and internal forces acting on the atomistic and bridging domain atoms, respectively. The derivation of Eq. (8) can be found in [5]. The resulting global stiffness matrix for the multiscale system, as shown in Eq. (8), will be a sparse symmetric matrix with off-diagonal components representing the nonlocal nature of the atomistic model and the interactions between atoms and the finite element nodes in the bridging domain. This system can then be solved iteratively using a nonlinear finite element solution procedure such as the Newton-Raphson method [5].

4.3. Temperature Effects

The solution to Eq. (8) provides the minimum potential energy configuration of the system. This minimum potential energy is analogous to carrying out a molecular statics simulation where temperature effects are essentially ignored. The use of MD can solve this issue as temperature effects are inherently considered through tracking of the atom velocities. Many approaches incorporate MD into the solution procedure; however, as discussed in Sec. 3, this requires small timesteps to accurately account for the atomistic vibrations. This timestep limits the model to very short simulation times as well as high strain rates, both of which are not practical for many engineering applications. To avoid this issue, the BCM has employed a temperature dependent potential into the energy formulation. One such potential energy function is the pairwise potential developed by Subramaniyan and Sun [52] where the interatomic energy takes into account the thermal expansion of the lattice constant. More recently the local quasiharmonic approximation [53, 54] has been used with the BCM [6], where the vibrational energy of the atoms about their minimum energy configuration is incorporated into the potential.

By solving the system using a temperature dependent potential, the BCM acts to minimize the free energy of the system as opposed to the potential energy. The result is an approach which allows one to solve isothermal problems under quasistatic loading conditions. If modeling more dynamic conditions is a necessity, the BCM equations could be solved by incorporating a dynamic finite element solver.

5. Performance Tests

5.1. Presence of Bridging Domain

To demonstrate the impact of the bridging domain weighting, a 1D example, as shown in Figure 3, is presented. The system consists of a chain of atoms where the atomistic domain is defined in the center with a section of the chain discretized for the bridging domain. The model symmetric chain consists of continuum elements on both ends, two bridging domain elements that contain four atoms each, and three atoms in the atomistic domain. The end of the chain was fixed while a force was applied to the opposite end. Each atom interacted with its nearest and second nearest neighbors based on a two-body harmonic potential, while the continuum elements and bridging cells were described by the Cauchy-Born theory [56]. Atom displacements in the bridging domain were mapped to the bridging cell nodes using mapping matrix **M**. This resulted in a reduction in the degrees of freedom to be solved from thirteen to nine.

The energy in the bridging domain was made up of the energy from the nonlocal atomic interactions within each cell as well as the weighted local energy based on the continuum Cauchy-Born theory. The weighting, calculated using Eq. (3), was multiplied by the local continuum energy stiffness matrix and added to the nonlocal stiffness matrix,

$$\mathbf{K}_B = \mathbf{w}^{(c)}\mathbf{K}_B^L + \mathbf{M}^T\mathbf{K}_B^{NL}\mathbf{M}. \tag{9}$$

As can be seen from the Figure 4, with no adjustment made in the bridging domain, the deformation is consistent throughout the continuum and atomistic domains; however it is highly nonlinear in the bridging domain. In this case only the second term in Eq. (9) is contributing to the stiffness of the bridging domain cell. Since atoms close to the continuum domain do not have a full set of neighboring atoms to interact with, the result will be inhomogeneous deformation. For the case where the local stiffness contribution is added to the bridging domain, the result is homogeneous deformation throughout the system, as also seen in Figure 4. Thus, the additional local stiffness being added to the bridging cell accounts for the missing nonlocal atomic interactions that result from the bridging/continuum domain interface.

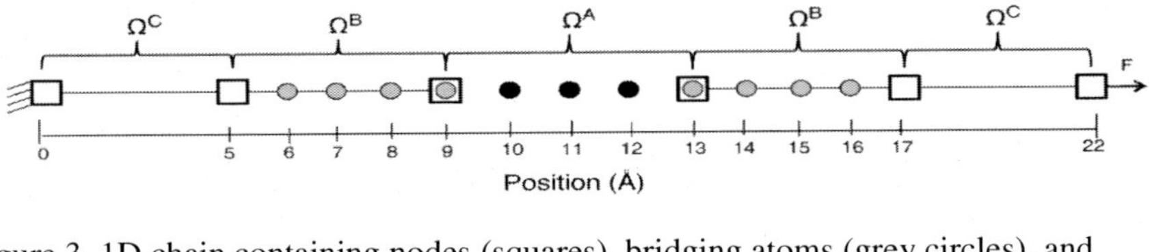

Figure 3. 1D chain containing nodes (squares), bridging atoms (grey circles), and atoms (black circles).

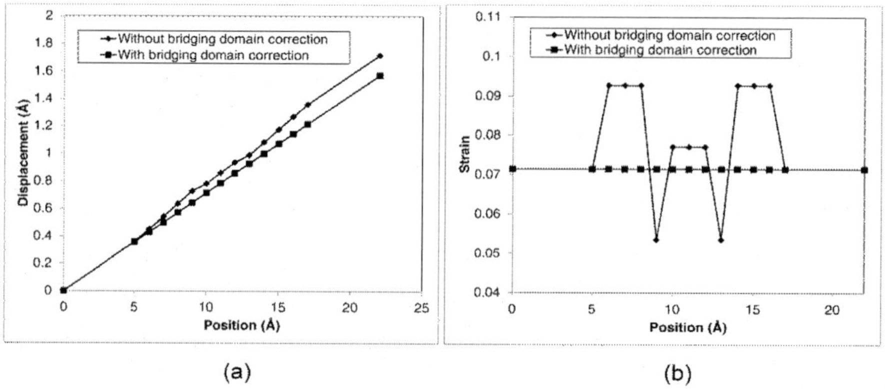

Figure 4. (a) Displacement and (b) strain with respect to initial position for a 1D chain with and without the bridging domain correction.

5.2. Comparison to Full MD

A number of case studies have been carried out [5-8] in order to gain insight into the effectiveness of the BCM. In these cases the comparison was made with a full MD simulation of the same system. The objective of the case studies was to see if the BCM could reduce the computational expense of a full MD simulation while reproducing the MD result for the observables of interest.

5.2.1. Stress Around a Void

Shown in Figure 5 is the multiscale model of a void in copper. The 2D model is of the (111) close-packed plane with biaxial loading applied in the x and y-directions.

An atomistic domain, containing 3229 atoms, surrounded the void, while the fully atomistic model consisted of 28833 atoms. The temperature dependent Lennard-Jones potential developed by Subramaniyan and Sun [52] was used in the multiscale model to account for the effects of temperature. For the full MD simulation, the timestep used was 1fs with the temperature held constant. The quasistatic loading was implemented by applying a biaxial strain to the specimen and then allowing the system to equilibrate at the applied strain.

At an applied strain of 3% the displacement in the atomistic region and the respective error between the BCM and MD models was calculated for temperatures 300K and 600K. The error for the 600K case was 7% and approximately 1.5 times that of the 300K case. As expected, at high temperature the displacement is larger as a result of a lower stiffness. As the temperature increased so too did the stresses and the displacements due to the material becoming less stiff as the atoms became more thermally excited.

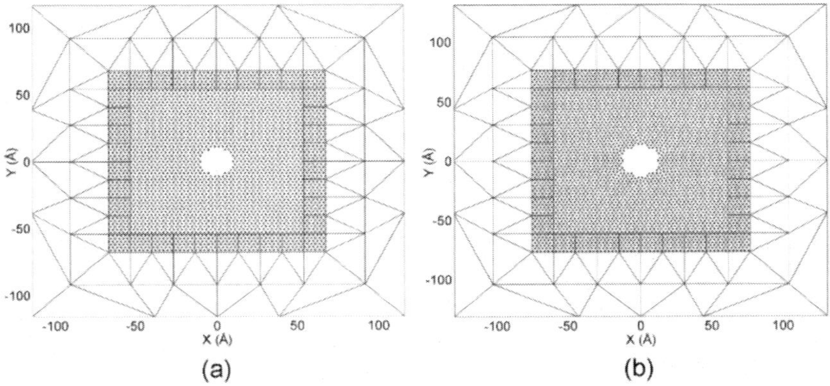

Figure 5. (a) Undeformed and (b) deformed bridging cell model of a nanovoid.

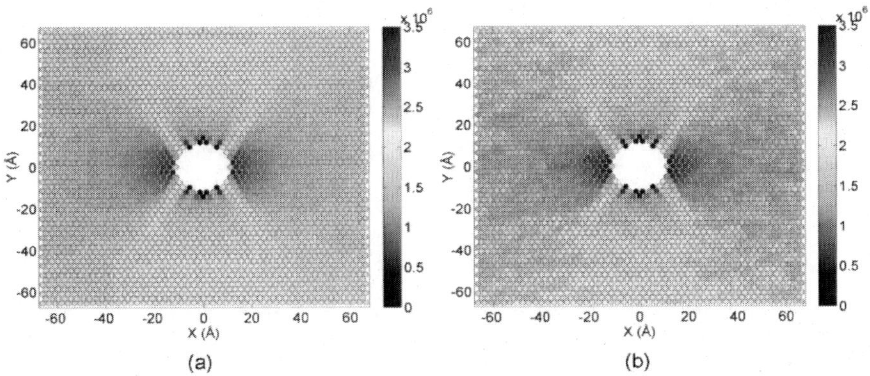

Figure 6. Y-component of atomic virial stress (kPa*Å^3) in bridging cell model for (a) 300K and (b) 600K.

The similarity between BCM and MD models was also seen through the stress plots of the atomistic and bridging domains, shown in Figure 6, for the Y-component of the stress at 600K. As can be seen from Figure 6, both the stress patterns and the stress values agree well between the two simulations. In general the stress patterns agree particularly well with what would be expected for void growth in fcc metals on the (111) plane. The highest stresses occurred around the void where the shear loop deformation had formed, indicating that the formation of cracks around the void in these regions would eventually lead to failure.

The smooth stress contours throughout the bridging domain at the edges of the plot also indicates that there is a seamless transition between atomistic and bridging domains. The fact that there are no abrupt changes in stress at the boundary indicates that the approximate linear mapping of atom displacements in this domain is accurate for the loading applied.

5.2.2. Nanoindentation

As presented in [7], the simulation of nanoindentation in copper was carried out using both a BCM and full MD simulation. The BCM model, as shown Figure 7, contained 6048 atoms while the full MD model was made up of 42856 atoms. The base of the model was fixed while the indenter was lowered at increments of 0.1Å to a depth of 11Å, allowing the system to relax after every load step to ensure the system remained stable.

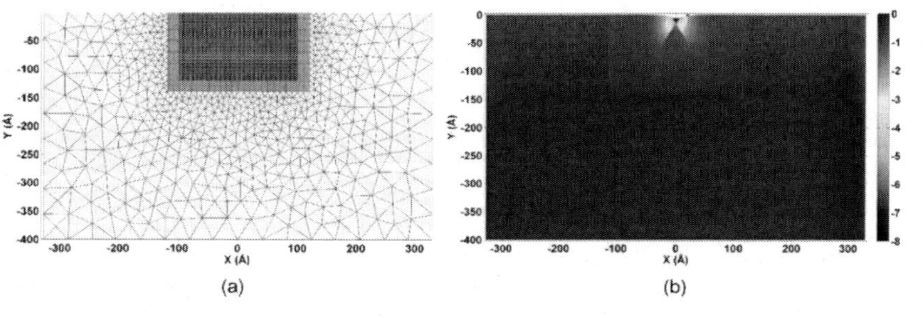

Figure 7. (a) BCM nanoindentation model and (b) resulting displacement plot [7].

The load vs displacement plots for the nanoindentation at different temperatures are shown in Figure 8. The increase in temperature resulted in a decrease in stiffness within the atomistic system as well as caused the onset of dislocations to occur at lower indentation depths. This behavior was observed for both BCM and full MD models. Comparing the two models with respect to the load at which the first dislocation forms, it was also observed that the difference between the two tended to increase with increasing temperature. This indicated the presence of an upper temperature limit where the temperature dependent potential is still applicable. Nevertheless, the multiscale model was able to reproduce the fully MD results at only 8% of computational time it took to run the full MD simulation.

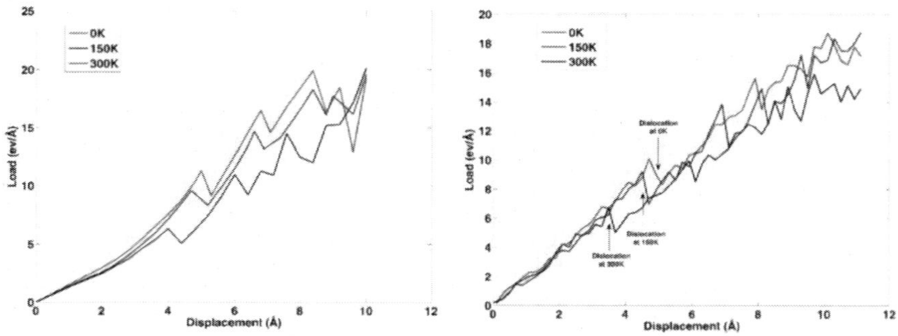

(a) (b)

Figure 8. Load vs displacement curves at different temperatures for (a) BCM model and (b) full MD model [7].

5.3. Extension of BCM to 3D Modeling

The BCM has recently been extended to modeling 3D deformation problems. In particular, the BCM has been applied to the modeling of 3D crack growth in gold and nickel single crystals under both mono and cyclic loading conditions [6, 8]. Two crystal orientations were studied; one in which brittle fracture was induced and another where crack tip blunting produced a ductile failure mode. Cyclic loading was applied by displacing the ends of the model in the z-direction to produce mode-I crack opening. This loading was applied in a cyclic fashion to produce progressive crack growth with each loading cycle. The resulting crack opening is shown for gold with two crystal orientations in Figure 9.

From the multiscale simulations the crack growth rate and stress intensity factor were calculated. These results were plotted, as shown in Figure 10, and the data was fit to the Paris law [57],

$$\frac{da}{dN} = C(\Delta K)^m \tag{10}$$

where da/dN is the crack growth rate, K is the stress intensity factor, and C and m are Paris parameters used for fitting the data.

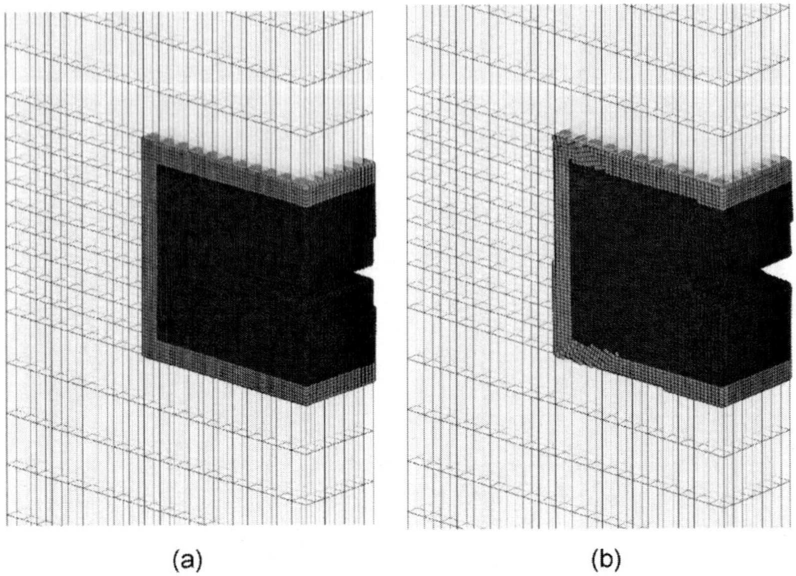

Figure 9. Crack opening and system deformation for (a) brittle crystal orientation and (b) ductile crystal orientation.

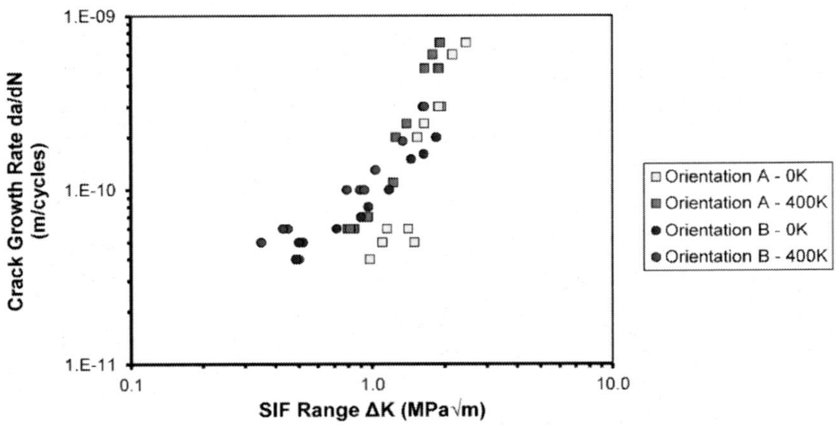

Figure 10. Crack growth rate vs stress intensity factor for crystal orientation A (brittle) and orientation B (ductile) of single crystal gold.

In general it was found that Paris parameter C tended to increase with increasing temperature while there was relatively little change in parameter m [6, 8]. Furthermore, increasing temperature saw a decrease in the threshold stress intensity factor, indicating that higher temperatures resulted in the onset of crack propagation occurring at a lower stress as well as faster crack propagation once the threshold stress was reached. This type of behavior agreed well with experimental results for single crystal metals [58, 59] and other MD based simulations of crack growth under cyclic loading [60].

5.4. Polymeric Systems

To demonstrate the generality of the BCM, the approach has also been applied to amorphous material structures. In the case of amorphous and in particular polymeric materials, different issues arise when coupling an atomistic to continuum domain. Since atoms in a polymer chain do not have a set configuration and may not adhere to affine deformations, the direct application of concurrent multiscale methods that resolve the mesh down to the atomic spacing are not suitable.

As discussed in Sec. 2, a number of multiscale methods exist that do not require the resolution of the finite element mesh down to the atomic scale, such as the bridging scale method [34], bridging domain method [23], atomistic-to-continuum (AtC) methods [30-33], and Arlequin method [24, 27]. Approaches based on these methods have also been recently applied to hybrid continuum-atomistic modeling of polymers; however these approaches involve coupling finite element and MD with small timesteps [61] or using a molecular statics simulation with no temperature considerations [62].

As stated in Sec. 4, due to the multi-body nature of polymeric energy functions, where bonding, angle and dihedral interactions between groups of atoms are considered, a strict mapping of the atom displacements in the bridging domain will produce high energy configurations as the polymer chains do not follow an affine mapping. As a result polymer chains in the bridging domain are allowed to relax once the local continuum and bridging domain residual is below a set tolerance. During this energy minimization, a number of atoms are designated as “support atoms” and are fixed within the bridging domain while the other atoms are free to move. These support atoms, as shown by the red atoms in Figure 11, are assigned at the beginning of the simulation to ensure that the polymer chains do not move completely outside of the bounds of the bridging domain.

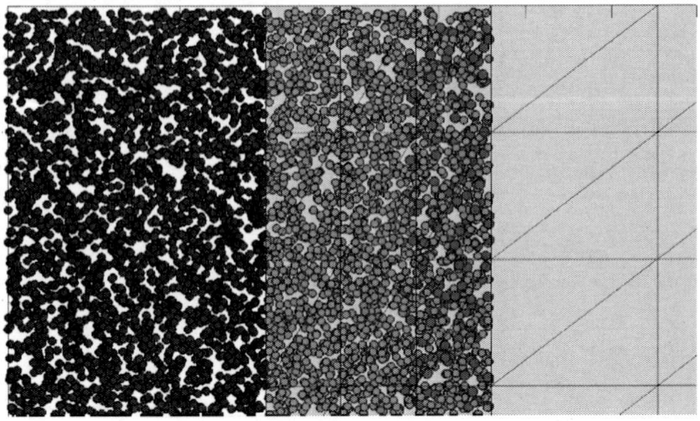

Figure 11. Fixed support atoms in the bridging domain.

In the polymeric version of the BCM, the temperature is incorporated into the coarse grained force field being used. In this sense the energy minimization taking place is that of the free energy of the system. The coarse grained potential is derived from a full atomistic simulation of the polymer chains based on a Boltzmann inversion [63]. During the iterative solution process, high energy configurations of polymer chains within the bridging domain are allowed to relax and are then remapped using an inverse isoparametric mapping procedure [64]. Thus far, coarse grained models of polyimide and polyethylene undergoing tensile deformation, as shown in Figure 12, have produced results for elastic modulus and stress that agreed well with those found in the literature [65].

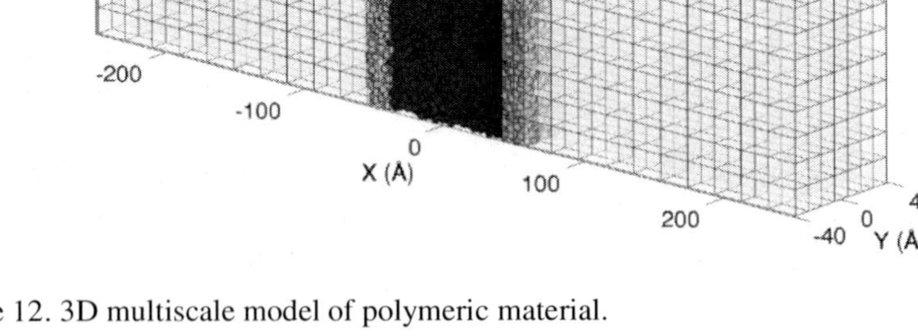

Figure 12. 3D multiscale model of polymeric material.

FUTURE DIRECTIONS

The application of multiscale modeling into more complex material systems such as polymeric and biomaterials will be necessary as the growth of nanotechnology extends into these fields. However, more research has to be done in order to achieve reliable computational methods to complement current experimental work.

The future growth of the BCM will look into expanded application spaces such as advanced material designs and structures where an ever growing emphasis on design from the nano- to the macro-level will require modeling across coupled length and time scales. For instance, in the field of biomechanics and biomaterials there have been significant advances in tissue engineering including the replacement and repair of damaged or diseased organs [66], tendons [67], and ligaments [68]. The recent growth in tissue engineering has resulted in a need for accurate models of both hard and soft tissue in order to create a cost effective method for designing new biomaterials for the medical field. Multiscale models of both hard and soft tissue could provide an improved understanding of failure of biological tissues, aid in the design of new biomaterials, and provide biomechanical insight into disease mechanisms.

On a separate front, nanocoating technology presents an advanced means of controlling material and structural properties at the nanoscale. Nanocoatings made up of as ceramic [69], metallic [70, 71], and organic materials [72] have proven to increase durability, strength, and resistance to wear and corrosion. Having advanced computing methods available to gain insight into the effect of various material parameters and properties would allow for accelerated development of nanocoating technology. The difficulty in using existing computational approaches is the multiscale nature of nanocoatings, thus making a method such as the BCM necessary for such an application. This logic extends to a number of emerging technologies such as nanosintering [73, 74], nanocomposites [75, 76], and nanometric machining [77, 78], which all involve material and structural behavior influenced by highly coupled length and time scales.

In order to meet this growing application space, current research on the BCM is being carried out with respect to expanded simulation sizes by means of large scale parallelization. Since the BCM has been formulated in a finite element framework, the use of advanced multi-core platforms running established finite element solvers could be implemented [79]. This would extend the applicability of the BCM to larger scale engineering problems on the micro- and macro-levels. In addition to parallelization, adaptive meshing of atomistic and continuum domains, such that the mesh evolves dynamically with the system, would further extend the applicability of the BCM. Extension of the BCM to modeling non-equilibrium temperature and thermal transfer problems would also open up many new applications. Incorporating the growth of dislocations from the atomistic domain into the continuum domain along with investigating the use of new mapping functions for the bridging domain is also an active area of research.

CONCLUSION

The future of multiscale modeling should lie in optimizing codes, generalizing the multiscale methodology, and demonstrating that multiscale methods can be applied to realistic system sizes and problems. By taking advantage of traditional finite element processes, the application of the BCM into large scale system sizes that are more applicable to modeling current problems in engineering will hopefully be realized. Its formulation also represents a generalized methodology that can be applied to a number of material structures as shown by its use in modeling both amorphous and crystalline material structures. The BCM has shown to be an effective means of coupling atomistic and continuum formulations; however, as discussed above, there are many areas where the technique can be further developed. Many exciting developments in multiscale modeling are expected in the coming years, with the continued growth of multiscale modeling playing an important role in the effort to produce and analyze new technological advances.

REFERENCES

[1] Clayton, J.D.; Chung, P.W. *J. Mech. Phys. Solids* 2006, 54(8), 1604-39.

[2] Ghosh, S.; Lee, K.; Moorthy, S. *Comput. Method. Appl.* M. 1996, 132 (1-2), 63-116.

[3] Fish, J.; Belsky, V. *Comput.Method. Appl.* M. 1995, 126(1-2), 1-16.

[4] Iacobellis, V.; Behdinan, K. *J. Appl. Mech.* 2013, 80(5), 051003.

[5] Iacobellis, V.; Behdinan, K. *Int. J. of Num. Meth. in Engrg.*, 2012, 92(7), 652-670.

[6] Iacobellis, V.; Behdinan, K. *Int. J. of Num. Meth. in Engrg.*, 2015, under review.

[7] Iacobellis, V.; Behdinan, K. TCMS, 2013, 2(6), 285-293.

[8] Iacobellis, V.; Behdinan, K. Proc. of the CSME 2014 Inter. Cong., June 2014.

[9] Tadmor, E.B.; Ortiz, M.; Phillips, R. *Philos. Mag. A* 1996, 73(6), 1529–1563.

[10] Eidel, B.; Stukowski, A. J. *Mech. Phys. Solids* 2008, 57(1), 87-108.

[11] Knap, J.; Ortiz, M. J. *Mech. Phys. Solids* 2001, 49(9), 1899–1923.

[12] Shenoy, V. B.; Miller, R.; Tadmor, E. B.; Rodney, D.; Phillips, R.; Ortiz, M. J. *Mech. Phys. Solids* 1999, 47(3), 611–642.

[13] Miller, R.; Tadmor, E.B.; Phillips, R.; Ortiz, M. Modell. *Simul. Mater. Sci. Eng.* 1998, 6(5), 607–638.

[14] Miller, R.; Tadmor, E.B. J. Comput. *Aided Mater. Des*. 2002, 9(3), 203–239.

[15] Shenoy, V.; Phillips, RMat. *Res. Soc. Symp. Proc*. 1999, 538, 465–471.

[16] Tang, Z.; Zhao, H.; Li, G.; Aluru, N. R. *Phys. Rev. B* 2006, 74, 0641101–16.

[17] Dupuy, L.M.; Tadmor, E.B.; Miller, R.E.; Phillips, R. *Phys. Rev. Lett.* 2005, 95(6), 060202.

[18] Kulkarni, Y.; Knap, J.; Ortiz, M. J. *Mech. Phys. Solids* 2008, 56(4), 1417–1449.

[19] Tadmor, E.B.; Legoll, F.; Kim, W.K.; Dupuy, L.M.; Miller, R.E. *Appl. Mech. Rev.* 2013, 65(1), 010803–010803.

[20] Broughton, J.; Abraham, F.; Bernstein, N.; Kaxiras, E. *Phys. Rev. B* 1999, 60(4), 2391–2403.

[21] Kohlhoff, S.; Gumbsch, P.; Fischmeister, H.F. *Philos. Mag. A* 1991, 64(4), 851–878.

[22] Shilkrot, L.E.; Miller, R.E.; Curtin, W.A. *Phys. Rev. Lett*. 2002, 89(2), 025501.

[23] Xiao, S.P.; Belytschko, T. *Comput. Methods Appl. Mech. Eng*. 2004, 193(17–20), 1645–1669.

[24] Dhia, H.B., Comptes Rendus de l'Académie des Sciences - Series IIB - Mechanics-Physics-Astronomy, 1998, 326(12), 899–904.

[25] Dhia, H.B., Rateau, G. *Int. J. Numer. Meth. Eng*. 2005, 62(11), 14421462.

[26] Xu, M.; Belytschko, T. Int. J. of Num. *Meth. in Engrg*. 2008, 76(3), 278-94.

[27] Gracie, R.; Belytschko, T. *Int. J. of Num. Meth. in Engrg*. 2009, 78(3), 354-78.

[28] Khare, R.; Mielke, S.L.; Schatz G.C.; Belytschko, T. Comput. *Methods Appl. M*. 2008, 197(41-42), 3190-202.

[29] Xu, M.; Gracie, R.; Belytschko, T. Int. J. of Num. *Meth. in Engrg*. 2010, 81(13), 1635-58.

[30] Badia, S.; Bochev, P.; Lehoucq, R.; Parks, L.M.; Fish, J.; Nuggehally, M.; Gunzburger, M. *Int. J. Multiscale Com*. 2007, 5(5), 387–406.

[31] Badia, S.; Parks, M.; Bochev, P., Gunzburger, M., Lehoucq, R. *Multiscale Model Sim*. 2008, 7(1), 381–406.

[32] Fish, J.; Nuggehally, M.A.; Shephard, M.S.; Picu, C.R.; Badia, S.; Parks, M.L.; Gunzburger, M. *Comput. Methods Appl. M.* 2007, 196(45-48), 4548–4560.

[33] Parks, M.L.; Bochev, P.B.; Lehoucq, R.B. *Multiscale Model. Sim.,* 2008, 7(1), 362–380.

[34] Wagner, G. J.; Liu, W. K. *J. Comput. Phys*. 2003, 190, 249–274.

[35] Karpov, E. G.; Yu, H.; Park, H.; Liu, W. K.; Wang, J. Qian, D. *Int. J. Solids Struct.* 2006, 43(21), 6359–6379.

[36] Qian, D.; Wagner, G. J.; Liu, W. K. *Comput. Meth. Appl. Mech. Eng*. 2004, 193, 1603–1632.

[37] Park, H. S.; Karpov, E. G.; Liu, W. K.; Klein, P. A. *Phil. Mag*. 2005, 85(1), 79–113.

[38] To, A. C.; Li, S. *Phys. Rev.B,* 2005, 72, 035414.

[39] Li, X.T.; E, W. J. *Mech. Phys. Solids* 2005, 53, 1650–1685.

[40] E, W.; Engquist, B. *Comm. Math. Sci.* 2003, 1(1), 87–132.

[41] Saether E.; Yamakov, V.; Glaessgen, E.H. *Int. J. Num. Meth. Engng*. 2009, 78, 1292-1319.

[42] Rudd, R. E.; Broughton, *J. Q. Phys. Rev. B, 1998*, 58, 5893–5896.

[43] E, W.; Li, X. Multiscale Modeling for crystalline solids. In: S. Yip (Ed.), Handbook of Multiscale Material Modeling, Springer, NY, 2005.

[44] Curtin, W. A.; Miller, R. E. Model. *Sim. Mater. Sci. Engng*. 2003, 11, 33–68.

[45] Miller, R.E.; Tadmor, E.B. Modell. *Simul. Mater. Sci. Eng*. 2009, 17(5), 053001.

[46] Dobson, M.; Luskin, M. *Math. Model. Anal.* 2008, 42(1), 113-39.

[47] Lin, P.; Shapeev, A.V. ArXiv E-Prints 2009, eprint arXiv:0909.5437.

[48] Shimokawa, T; Mortensen, J.J.; Schiøtz, J.; Jacobsen, K.W. *Phys. Rev. B* 2004, 69(21), 2141041-10.

[49] E, W.; Lu, J.; Yang, J.Z. *Phys. Rev. B* 2006, 74(21), 2141151-12.

[50] Omelyan, I.; Kovalenko, A. *J. Chem. Phys.* 2013, 139(24), 244106.
[51] Vinogradov, O. *Comput. Mater. Sci.* 2008, 41(4), 493-7.
[52] Subramaniyan, A.K.; Sun, C.T. *Nanotech.*, 2008, 19(28), 285706.
[53] Sutton, A.P. *Philos. Mag.* A 1989, 60(2), 147-59.
[54] Lesar, R.; Najafabadi, R.; Srolovitz, D.J. *Phys. Rev. Let.* 1989, 63(6), 624-7.
[55] Liu, B.; Huang, Y.; Jiang, H.; Qu, S.; Hwang, K.C. Comput. *Methods Appl. M.* 2004, 193(17-20), 1849-64.
[56] Zanzotto, G. *Acta Crysta. Sec. A* 1996, 52(6), 839-49.
[57] [Paris, P.C.; Erdogan, F. *J. Basic Engng*. 1960, 85, 528-534.
[58] Liaw, P.K.; Lea, T.R., Logsdon, W.A. *Acta Metal.*1983, 31(10), 1581-1587.
[59] Hicks, M.A.; King, J.E. *Inter. J.of Fatigue* 1983, 5(2), 67-74.
[60] Potirniche, G.P.; Horstemeyer, M.F.; Gullett, P.M., Jelinek, B. *Proc. of the Royal Society A* 2006, 462(2076), 3707-3731.
[61] Pfaller, S.; Rahimi, M.; Possart, G.; Steinmann, P.; Müller-Plathe, F.; Böhm, M.C. *Comput. Methods Appl. Mech. Eng*. 2013, 260, 109-29.
[62] Bauman, P.T.; Oden, J.T.; Prudhomme, S. *Comput. Methods Appl. Mech. Eng*. 2009, 198(5-8), 799-818.
[63] Reith, D.; Pütz, M.; Müller-Plathe, F. *J. Comput. Chem.* 2003, 24(13), 1624-36.
[64] Chinnaswamy, C.; Amadei, B.; Illangasekare, T.H. *Int. J. Numer. Methods Eng*. 1991, 31(7), 1253-70.
[65] Hergenrother, P.M.; Watson, K.A.; Smith Jr., J.G.; Connell, J.W.; Yokota, R. *Polymer* 2002, 43(19), 5077-93.
[66] Bishop, J. E. Mol. *Med. Today* 1998, 4(2), 69-75.
[67] Woo, S.L.; Hildebrand, K.; Watanabe, N.; Fenwick, J.A.; Papageorgiou, C.D.; Wang J.H. *Clin. Orthop. Relat. R.* 1999, 367, S312-23.
[68] Qian, D.; Liu, W.K. *Comput. Mech*, 2008, 42(4), 483-4.
[69] Hu, X.; Shen, H.; Cheng, Y.; Xiong, X.; Wang, S.; Fang, J.; Wei, S. *Surf. Coat. Tech.* 2010, 205(7), 2000-6.
[70] Ivosevic, M.; Knight, R.; Kalidindi, S.R.; Palmese, G.R.; Sutter, J.K. High Perform. *Polymers* 2003, 15(4), 503-17.
[71] Robitaille, F.; Yandouzi, M.; Hind, S.; Jodoin, B. *Surf. Coat. Tech.* 2009, 203(19), 2954-60.
[72] Martin, T.P.; Lau, K.K.S.; Chan, K.; Mao, Y.; Gupta, M.; O'Shaughnessy, W.S.; Gleason, K.K. *Surf. Coat. Tech.* 2007, 201(22-23), 9400-5.
[73] Groza, J.R. *Nanostruct. Mater.* 1999, 12(5), 987-92.

[74] Fang, Z.Z.; Wang, H. *Inter. Mater. Rev*. 2008, 53(6), 326-52.

[75] Baur, J.; Silverman, E. *MRS Bulletin* 2007, 32(4), 328-34.

[76] Breton, Y.; Désarmot, G.; Salvetat, J.P.; Delpeux, S.; Sinturel, C.; Béguin, F.; Bonnamy, S. *Carbon* 2004, 42(5-6), 1027-30.

[77] Pei, Q.X.; Lu, C.; Fang, F.Z.; Wu, H. *Comput.Mater. Sci.* 2006, 37(4), 434-41.

[78] Fang, F.Z.; Wu, H.; Zhou, W.; Hu, X.T. *J. Mater. Process Technol.* 2007, 184(1-3), 407-10.

[79] Belytschko, T.; Liu, W.K.; Moran, B. Nonlinear Finite Elements for Continua and Structures. John Wiley & Sons Ltd., West Sussex, England 2000.

In: Molecular Mechanics and Modeling
Editor: Priscilla Watkins
ISBN: 978-1-63483-388-2

Chapter 2

BASIC PROPERTIES OF A MOLECULAR MECHANICS PROGRAM AND THE GENERATION OF UNKNOWN STEREO STRUCTURES OF PROTEINS FOR QUANTITATIVE ANALYSIS OF ENZYME REACTIONS

Toshihiko Hanai*
Health Research Foundation,
Research Institute for Production Development 4F,
Sakyo-ku, Kyoto, Japan

ABSTRACT

Experimental results regarding the contribution of alkyl chains towards hydrogen-bonding capability of hydroxyl groups were analyzed using different versions of the MM2 program. The contribution of hydrogen-bonding energy values calculated using the MM2 program of the current CAChe programs differed significantly from those of previous ones, and the dihedral angle energy value was found to affect the contribution of alkyl chain length. The current report elaborates a step-wise process for inferring the stereo structure of proteins constructed

* Email: thanai@kf7.so-net.ne.jp.

using related stereo structures in spite of a less than 35% amino acid sequence similarity. A quantitative analysis of enzyme reactivity was performed on proteins such as a mutant of R-amino acid oxidase M215R and aromatic alcohol $NADP^+$ oxidoreductase ELIS-2.

INTRODUCTION

Molecular engineering approaches for the generation of improved molecules for applications in a variety of industrial, research, and medical settings are in increasing demand. Furthermore, specific techniques are used for establishing the biochemical identity as well as for modifying the original molecules. In this context, structure-based drug design, which specifically requires designed proteins, and advances in structural proteomics and computational modeling may eventually provide principals for the correlating the gene sequence with protein structure and function, so that the rational redesign of proteins is more feasible [1]. In this presentation, we describe a method for modification without lose of reaction properties. The protein structures thus constructed were optimized using the molecular mechanics program MM2. The current version of MM2 showed conformational differences, which were investigated using a pair of identical alkyl alcohols using alkanes as reference compounds. Alkyl chains up-to the butyl group are known to contribute towards hydrogen bonding, whereas longer alkyl groups do not affect the hydrogen bonding capability. These experimental results were analyzed using the different versions of the MM2 programs [2]. Hydrogen-bonding energy was calculated using MM2 program of the current CAChe.

Quantitative methods for the analysis of enzyme reactivity are critical for developing precise diagnostic methods and for drug discovery, and to reduce the number of tests on animals. Although the specific affinity of an enzyme reaction can be analyzed *in silico*, it is difficult to generalize the selectivity of an enzyme reaction.

Auto docking programs are not satisfactory for this purpose mainly because they recognize only the molecular shape and not the priority of the reaction center. Enzymatic reactions are typically simple chemical reactions that occur in specific chambers where electron transfer can occur under low energy conditions. One possible approach is the replacement of the original substrate used for obtaining a protein crystal with a new substrate, followed by quantitative analysis of the novel conformations produced.

This report demonstrates the contribution of hydrogen bonding and helps to infer the stereo structure of a protein by using a related stereo structure and the sequence information. Furthermore, it provides a method for replacement of a small substrate with a larger one, using the MM2 program for quantitative analysis of the protein and its substrate interactions. This performance of this method is demonstrated by quantitative analysis of the enzyme reactivity.

EXPERIMENTAL

PowerMAC 7000, PowerMac 8000, a Latitude 840, and Foxconn H61MX computers were used for CAChe™ version 3,7, 4.5, 5.0, 6.1, and 7.5 computational chemical programs, respectively. CAChe programs were purchased from Sony-Tektronix (Tokyo) and Fujitsu (Tokyo).

RESULTS AND DISCUSSION

1. Basic Study of Hydrogen Bonding

The hydrogen bonding capacity of alkyl alcohols depends on the alkyl chain length and up to four methylene units can affect this capacity. Here, a quantitative study of such experimental results was performed. Stereo structures ranging from methyl alcohol to decyl alcohol were constructed, and individual and identical pair energy values were obtained after optimizing the structures by MM2 calculations. The calculated energy values using the different versions of MM2 programs are summarized in Table 1, wherein the values obtained using CAChe versions 5 and 7.5 are enlisted. Calculated values obtained using versions 3.7, 4.5, and 5.0 were identical, and those obtained using versions 6.1, 7.5 and Scigress® 2.0 were identical. Although molecular interaction (MI) energy values calculated using the two groups were different, the atomic properties obtained using these MM2 programs appeared to be the same. The energy values of alkanes, especially the van der Waals energy values that indicate molecular size effects, were also calculated for reference. The optimized stereo structures of a pair of methyl alcohol, butyl alcohol, and octyl alcohol are shown in Figure 1a-f. Figures 1a-c were optimized using version 5 of the program, whereas Figures 1d-f were optimized using version 7.5.

Table 1. Molecular properties of a pair of alkylalcohols

	Version 3.7				Version 4.5			
	pfs	phb	pes	pvw	pfs	phb	pes	pvw
Methyl alcohol	-3.693	-2.897	-1.213	0.130	-3.689	-2.892	-1.211	0.128
Ethyl alcohol	-2.244	-2.898	-1.246	1.229	-2.244	-2.817	-1.187	1.023
Propyl alcohol	0.253	-0.137	-0.074	-0.361	0.253	-0.071	-0.076	-0.401
Butyl alcohol	0.602	-0.101	-0.051	-0.430	0.605	-0.079	-0.052	-0.443
Pentyl alcohol	0.919	-0.077	-0.068	-0.477	0.920	-0.079	-0.070	-0.473
Hexyl alcohol	1.207	-0.087	-0.056	-0.562	1.209	-0.090	-0.057	-0.555
Heptyl alcohol	1.497	-0.082	-0.062	-0.624	1.494	-0.080	-0.062	-0.627
Octyl alcohol	1.779	-0.079	-0.056	-0.704	1.783	-0.080	-0.058	-0.699
	Version 5.0				Version 5.0			
	pfs	phb	pes	pvw	dhang	ang	str	atrben
Methyl alcohol	-3.689	-2.892	-1.211	0.126	0	0.272	0.060	0
Ethyl alcohol	-2.417	-2.878	-1.239	1.190	0.073	0.351	0.054	0.031
Propyl alcohol	0.249	-0.070	-0.070	-0.406	0.137	0.510	0.163	0.070
Butyl alcohol	0.609	-0.074	-0.052	-0.439	1.419	0.739	0.288	0.115
Pentyl alcohol	0.925	-0.082	-0.071	-0.467	1.831	1.016	0.461	0.170
Hexyl alcohol	1.209	-0.083	-0.057	-0.560	1.085	1.222	0.541	0.204
Heptyl alcohol	1.498	-0.081	-0.067	-0.612	1.888	1.133	0.514	0.194
Octyl alcohol	1.779	-0.090	-0.058	-0.697	1.111	1.406	0.671	0.241
Nonyl alcohol	2.080	-0.076	-0.064	-0.757	0.088	1.599	0.790	0.283
Decyl alcohol	2.362	-0.080	-0.058	-0.834	0.094	1.685	0.885	0.312
	Version 6.1							
	Pfs	phb	pes	pvw				
Methyl alcohol	-6.136	-5.136	-0.988	-0.452				
Ethyl alcohol	-5.043	-5.126	-0.988	0.461				
Propyl alcohol	-3.908	-5.117	-0.944	1.148				
Butyl alcohol	-2.477	-5.067	-0.538	0.959				
Pentyl alcohol	-1.819	-5.039	-0.461	0.630				
Hexyl alcohol	-1.488	-5.014	-0.345	0.480				
Heptyl alcohol	-1.236	-5.032	-0.428	0.375				
Octyl alcohol	-0.910	-5.014	-0.345	0.338				
Nonyl alcohol	-0.664	-5.025	-0.390	0.245				
Decyl alcohol	-0.320	-5.012	-0.337	0.212				
	Version 7.5				Version 7.5			
	pfs	phb	pes	pvw	dhang	ang	str	strben
Methyl alcohol	-6.293	-5.136	-0.995	-0.448	0.007	0.272	-	-
Ethyl alcohol	-5.033	-5.123	-0.970	0.438	0.181	0.345	0.045	0.031
Propyl alcohol	-3.910	-5.116	-0.943	1.148	0.273	0.507	0.155	0.069
Butyl alcohol	-2.472	-5.062	-0.516	0.911	1.021	0.757	0.277	0.115
Pentyl alcohol	-1.820	-5.039	-0.465	0.628	1.446	1.029	0.414	0.164
Hexyl alcohol	-1.495	-5.023	-0.379	0.507	1.402	1.240	0.544	0.205
Heptyl alcohol	-1.239	-5.027	-0.410	0.361	1.516	1.420	0.668	0.244
Octyl alcohol	-0.926	-5.014	-0.353	0.324	1.425	1.606	0.796	0.284
Nonul alcohol	-0.690	-5.025	-0.389	0.222	1.475	1.781	0.927	0.323
Decyl alcohol	-0.357	-5.012	-0.334	0.174	1.435	1.978	1.042	0.362

pfs: pair of final structure; phb: pair of hydrogen bond; pes: pair of electrostatic; pvw: pair of van der Waals; dhand: dihedral angle; ang: angle; str: stretch; strben: stretch bend (unit: kcal/mol).

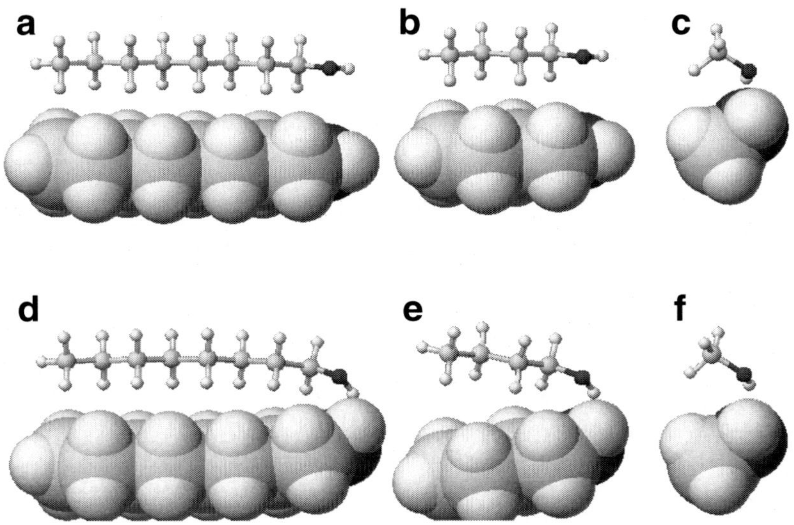

Figure 1. Complexes of pair of alkylalcohols optimized using version 5 and 7.5 MM2 programs: a & d: octylalcohoil; b & e: butylalcohol; c & e: methylalcohol.

The final structure energy (fs) for a pair of identical alcohols was less than two times the fs of a single alcohol, indicating that the energy values calculated using the MM2 program can explain the degree of molecular interaction The final structure energy is a combination of hydrogen bonding (hb), electrostatic (es), van der Waals (vw), bond stretch, bond angle dihedral angle, and improper torsion energies. The electrostatic, hydrogen bonding, and van der Waals energies are measures of ion-ion interaction levels hydrogen bonding contribution, and hydrophobic interaction, respectively. Further, steric hindrance is contributed by other energies.

The molecular conformations optimized using the latest MM2 program indicated a strong hydrogen bonding contribution from butyl alcohol and octyl alcohol hydroxyl groups. Compared to the hydroxyl groups optimized using the older version of MM2 program, the latest version indicated that they were bent and presented contacts as shown in Figure 1. The final (optimized) structure energy change from methyl alcohol to decyl alcohol was about 6 kcal/mol. A significant difference was observed in the hydrogen bonding energy values, where the energy values calculated using the previous and latest versions ranged between -3.7 to 2.4 and -5.1 to -5.0 kcal/mol, respectively. This indicated that the energy values calculated using the recent version were significantly lower. In addition, the contribution of electrostatic energy (~ 0.66 kcal/mol), calculated using the recent MM2 program version, was lower in comparison to that calculated using the older version (~ 1.2 kcal/mol). Furthermore, the van der Waals energy contribution, as calculated by the recent version, was also markedly lower. A significant difference in the dihedral angle energy was observed, which changed about 1.4 kcal/mol from methyl alcohol to decyl alcohol using the recent version. However, changes in the calculated energy using the older version did not increase consistently, and the values of methyl alcohol and decyl alcohol were approximately similar. The contributions from angle, stretch, and stretch bent energies calculated using both the older and the recent versions were highly similar. These results indicated that the atomic properties using both version of the program remained the same, but the molecules were more flexible in the recent version of the program. These results may contribute towards building the stereo structure of proteins.

2. Construction of Stereo Structures of Proteins

The quantitative analysis of an enzyme reaction is the basic requirement for the development of precise diagnostic methods, drug discovery processes, and for designing engineering enzymes. Calculating the precise contribution of hydrogen bonding energy, which is an important interaction force in biological systems, may improve the precision of protein stereo structures optimized using the MM2 program. However, handling the stereo structures using BioCAChe within version 5 of CAChe was tedious and it was difficult to demonstrate the conformations. To overcome this problem, the version 6 BioCAChe program with additional capabilities was used to analyze the enzyme reactivity of a mutant protein of yeast R-amino acid oxidase (DAO). The mutant 1C0P M215R was developed especially for analyzing the freshness of food [3]. The structure of yeast DAO, 1C0P was downloaded from Protein Data Bank [4], and the 215th methionine residue was replaced with arginine and the stereo structure was optimized using MM2 calculation. Furthermore, the original substrate was replaced by using the super impose method within the computational chemical program with several R-amino acids following which the new complex conformation was also optimized. The atomic distance between key atoms at the reaction site indicated reactivity of this enzyme reactivity with several R-amino acids. Furthermore, coenzyme and substitutes within 3Å from the substrate were extracted, locked, and atomic partial charges on these molecules were calculated using PM5 [5] of the molecular orbital package (MOPAC) program. The atomic partial charge based on the localization of electrons is an important factor for studying molecular interactions and reactivity of enzymes. The difference in atomic partial charge between the original structure and the complex demonstrated a good agreement with the enzyme reactivity [3]. In case of 1C0P M215R mutant, only one amino acid replacement was required. Hence, no further investigations are needed to create this new mutant.

Since the structure downloaded from Protein Data Bank has an unacceptable conformation for further studies using MM2 calculation. Few processes that need to be followed in order to perform the above experiment are as follows:

1. Download a selected file from PDB [4].
2. Valance the structure in order to add hydrogen.
3. Select water molecules and delete them.
4. For analyzing the contribution of water molecules in the enzyme reaction, select the substrate, select atoms within 3Å from the substrate, and further analyze whether water molecules exist or not. Water molecules were absent from the reaction center in case of the R-amino acid oxidase. This would ensure that the active site of the enzyme provides for a specific nonpolar environment with specific polar groups that are designed for efficient electrostatic stabilization of ionic structures in comparison with water or other solvents [6].
5. Observe and fix the bond length and conformation of protein, coenzyme, and substrate.
6. Finally, initiate a MM2 calculation to fix the conformation.
7. The enzyme reactivity can now be analyzed.

The replacement of few amino acids and construction of mutants is a simple task. For example, one amino acid can be replaced and the structure can be optimized in a manner similar to that for the 1C0P M215R mutant. In case a three amino acid replacement is desired, the same process must be repeated thrice. However, the replacement of several amino acids is not as simple as a single amino acid replacement. For this purpose, amino acid replacement has to be carefully ordered to avoid damage to the stereo structure. Once all amino acids were replaced, the structure of the resulting mutant was optimized using MM2. However, this optimized structure could not use for studying the enzyme reaction due to the damage caused to the original stereo structure.

During stereo replacement, all amino acids were first locked in order to avoid effects on the original stereo structure. Subsequently, selected amino acids at the beginning and end position of the protein were replaced under unlocked condition because changes to these amino acids do not damage the stereo structure of the protein. The resulting new structure was then optimized using the MM2 calculation.

In the second step, an amino acid within a helix was selected and replaced to a desired amino acid. The whole structure was optimized, while the new amino acid was kept unlocked. Further, the conformation of the replaced amino acid was analyzed to make sure that this amino acid was independent from other amino acids. Consequently, the newly replaced amino acid and both side amino acids were unlocked and the protein structure was optimized. The entire stereo structure was optimized again by unlocking all amino acids. Amino acids present in helices were unlocked, if necessary, before final optimization using MM2. This approach is demonstrated in the following step-wise directions.

1. Firstly, water oxygen atoms is shown as black dots in the stereo structure downloaded from the PDB file, as indicated in Figure 2.
2. Next, the protein structure is shown in wire conformation as seen in Figure 3. The coenzyme, flavin adenine dinucleotide (FAD), and the substrate, R-alanine, appear with unexpected molecules that are circled. Black dots represent water oxygen atoms.
3. Atoms and molecules are shown in Figure 2 after valancing, which adds hydrogen atoms. The resulting structure now appears complicated (Figure 4).
4. The protein structure is shown in wire conformation to clearly identify the position of water molecules (Figure 5).
5. Water molecules and unexpected molecules shown in circles are selected and erased. (Figure 6).
6. Very long bonds, which may indicate lysine ε-amino groups, are cut and the atoms are fixed (Figure 7).
7. Correct the flavin adenine dinucleotide (FAD) structure considering the p*K*a values of FAD. Carefully observe the structure of the phosphate moiety, especially the oxygen atoms therein. In case of R-amino acid oxidase, double bonds of aromatic rings do not appear to be appropriate. A similar observation was made with other coenzymes. As seen in Figure 8, the existence of one oxygen atom, or an overlap of two oxygen atoms, or the overlap of an oxygen atom

with a phosphine is unclear. In this case, erase one oxygen atom and correct the structure. In another case, one oxygen atom seems to overlap with a phosphine atom.

8. Correct the structure of R-alanine substitute. Alanine should be ionized due to the p*K*a values. At this stage, carefully observe the atomic orbitals since SP2 atoms may be indicated as SP3 atoms, for example, the carbon atom of the carboxyl group. Following this, optimize the total structure by using the MM2 calculation, as shown in Figure 9.
9. The structure has to be fixed with extra care if the following messages regarding irregular bonds and conformations are observed.

9a [This sample contains with non-standard hybridization or unsatisfied valance. Do you want to highlight them instead of saving?]

9b [No configurations in MolStruct file. Program stopped.]

Construction of a Stereo Structure of the D-Amino Acid Oxidase Mutant M215R

1. Replace the 215th methionine with arginine on the sequence table and optimized the structure. The location of R215 is shown in Figure 10.
2. The structure of R-amino acid oxidase 1C0P with R-alanine optimized using the MM2 program is shown on Figure 11.
3. The structure of 1C0P mutant M215R with R-alanine optimized using MM2 calculation is shown in Figure 12. This structure is identical to that in Figure 11, albeit displaying slight differences with respect to 1C0L. The structure of 1C0L optimized using the MM2 calculation with R-alanine, the methyl group of which faces the front, as shown in Figure 13.

Construction of a Stereo Structure of H. Sapiens R -Amino Acid Oxidase

Study of the Enzyme Reactivity of M215R R-Amino Acid Oxidase

Figure 16 shows the conformation around R-alanine in the structure extracted from Figure 12. Only selected amino acid residues are shown to demonstrate the reactivity of this enzyme. The carboxyl group oxygen atoms of alanine are in tight contact with the guanidyl group hydrogen atoms of arginine and the atomic distance is 2Å. The ionized amino group is located at the top of the oxygen atom of carbonyl group. The distance is 2.1Å.

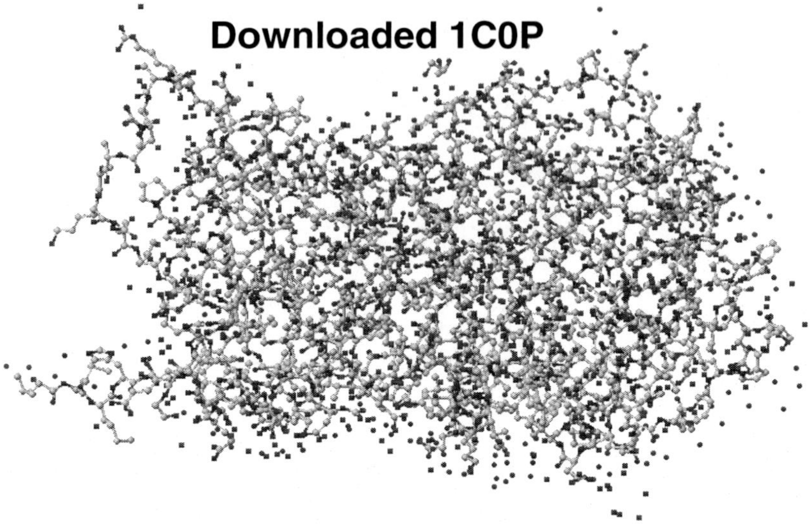

Figure 2. Downloaded stereo structure of R-amino acid oxidase 1C0P, Black dots: oxygen of water molecules. Light gray ball: carbon, small white ball: hydrogen, dark gray ball: nitrogen, black ball: oxygen, large black ball: phosphate.

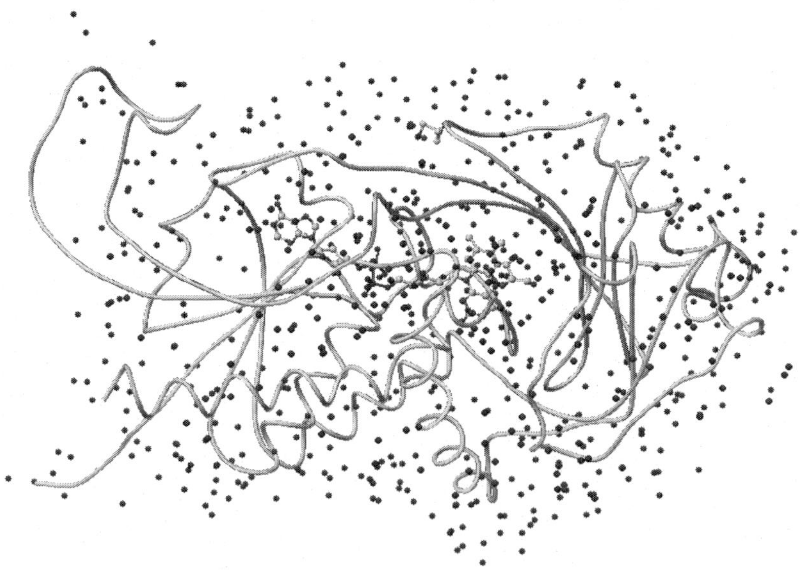

Figure 3. Downloaded stereo structure of R-amino acid oxidase 1C0P. Protein is shown as a wire flame form. Black dots: oxygen of water molecules. Symbols: see Figure 2.

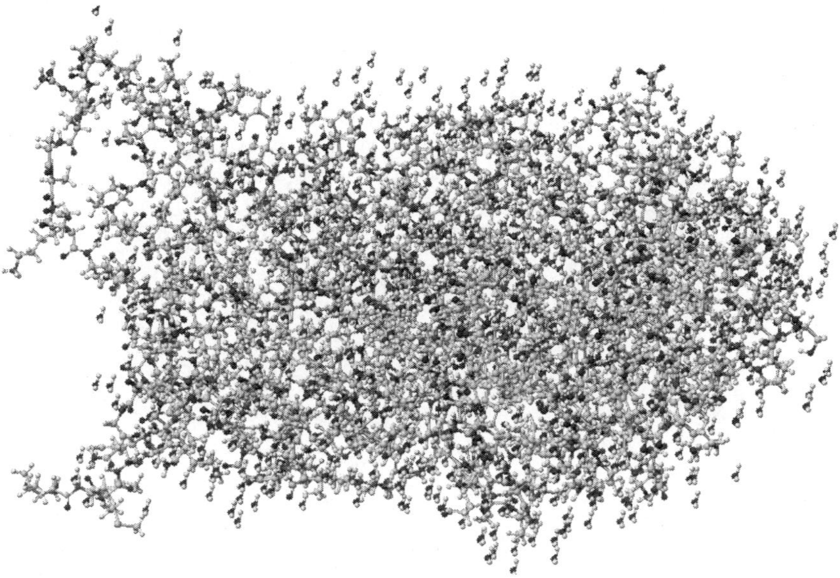

Figure 4. Added hydrogen atoms onto Figure 2 structure. Symbols: see Figure 2.

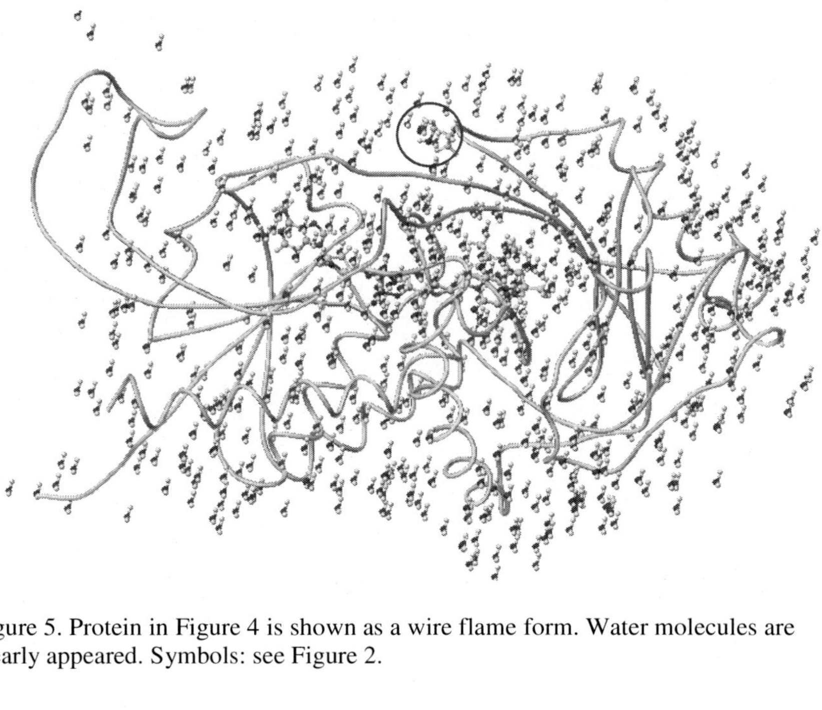

Figure 5. Protein in Figure 4 is shown as a wire flame form. Water molecules are clearly appeared. Symbols: see Figure 2.

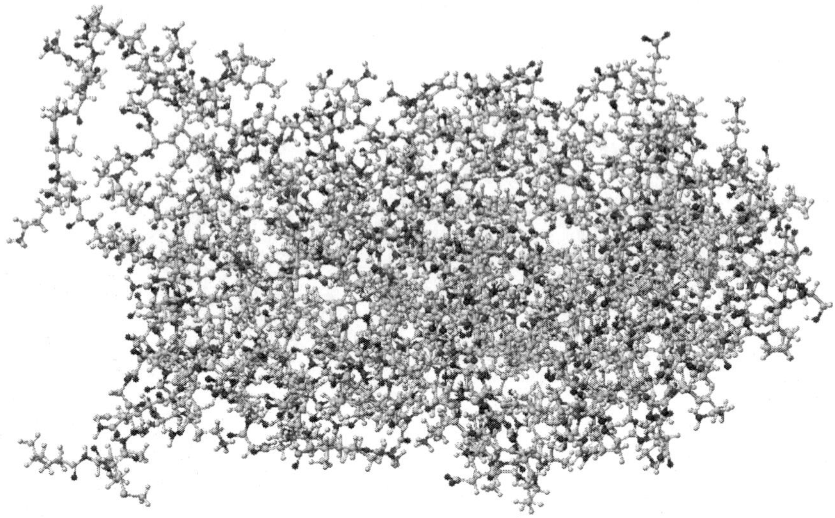

Figure 6. Erase water molecules and strange molecules. Symbols: see Figure 2.

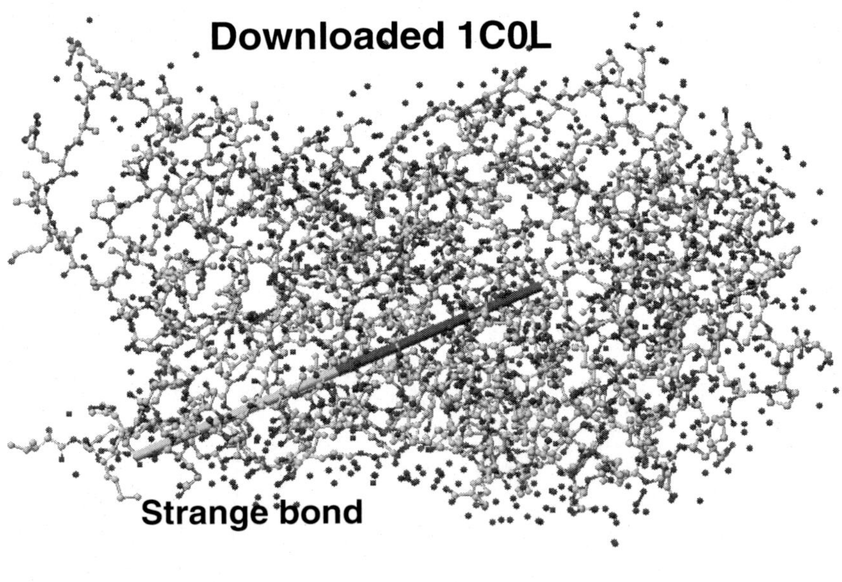

Figure 7. Strange bond in downloaded R-amino acid oxidase 1C0L The bond is indicated as 0.5Å cylinder instead of generally used 0.1Å. Symbols: see Figure 2.

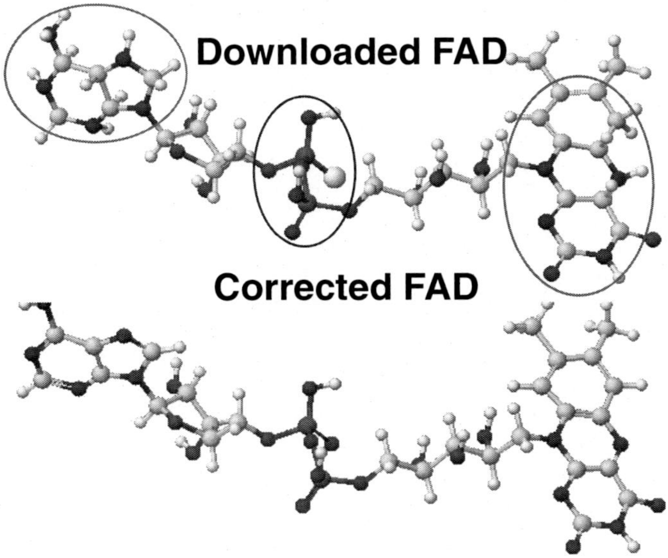

Figure 8. Extracted a downloaded and the corrected form coenzyme FAD. Strange atoms are circled. Symbols: see Figure 2.

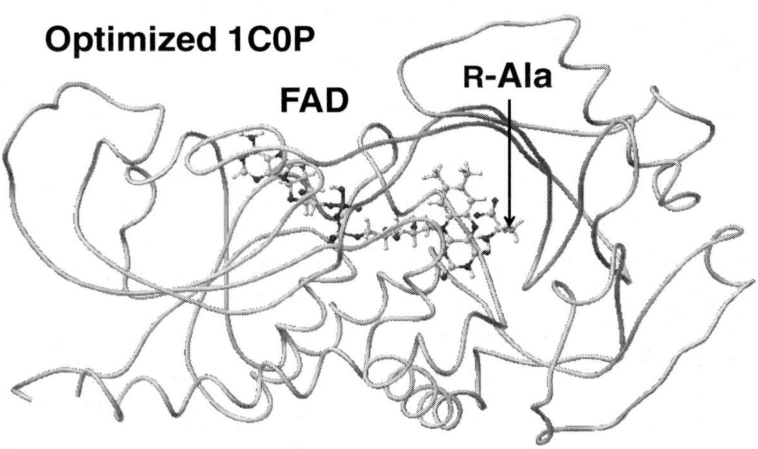

Figure 9. Corrected and optimized 1C0P structure using MM2 calculation. Protein is shown as a wire flame form. Symbols: see Figure 2.

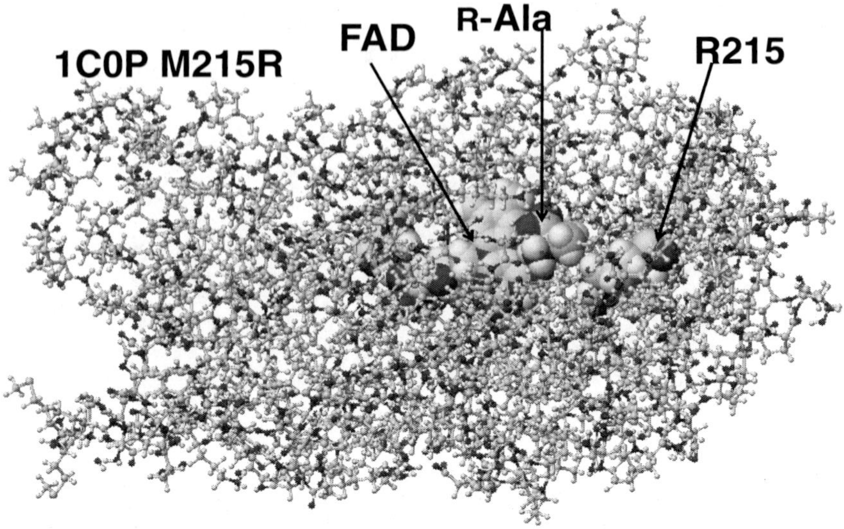

Figure 10. Indicates the location of 215th arginine with FAD and R-alanine. Symbols: see Figure 2. Atom size of R215, FAD, and R-alanine molecules is shown as their full size.

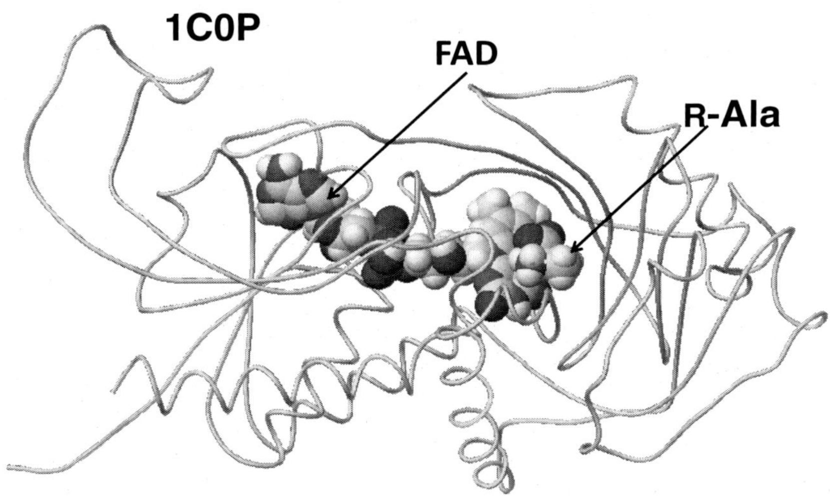

Figure 11. Optimized 1C0P structure shown as a wire flame form with FAD and R-alanine using MM2 calculation. Symbols: see Figure 2. Atom size of FAD, and R-alanine molecules is shown as their full size.

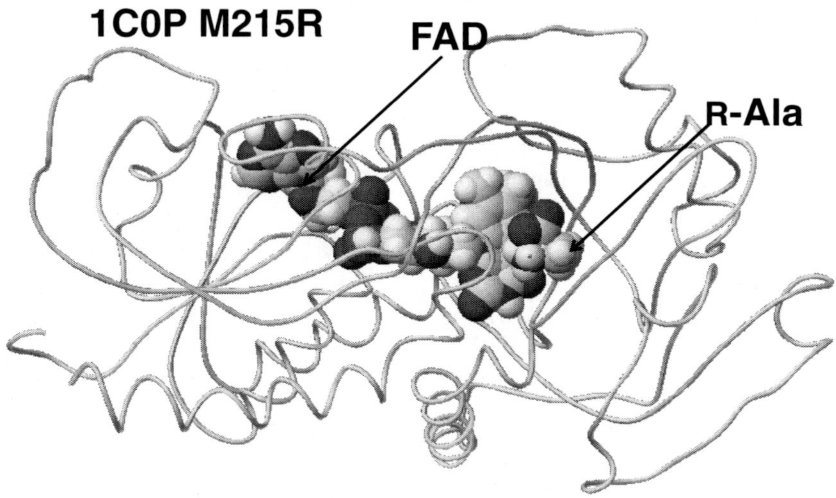

Figure 12. Optimized 1C0P mutant M215R as a wire flame form with FAD and R-alanine using MM2 calculation. Symbols: see Figure 2. Atom size of FAD, and R-alanine molecules is shown as their full size.

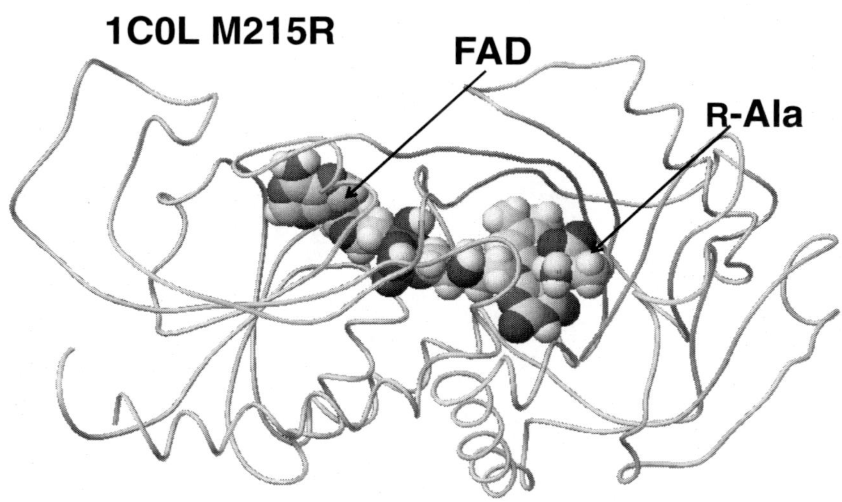

Figure 13. Optimized 1C0L mutant M215R as a wire flame form with FAD and R-alanine using MM2 calculation. Methyl group of alanine is faced to front. Symbols: see Figure 2. Atom size of FAD, and R-alanine molecules is shown as their full size.

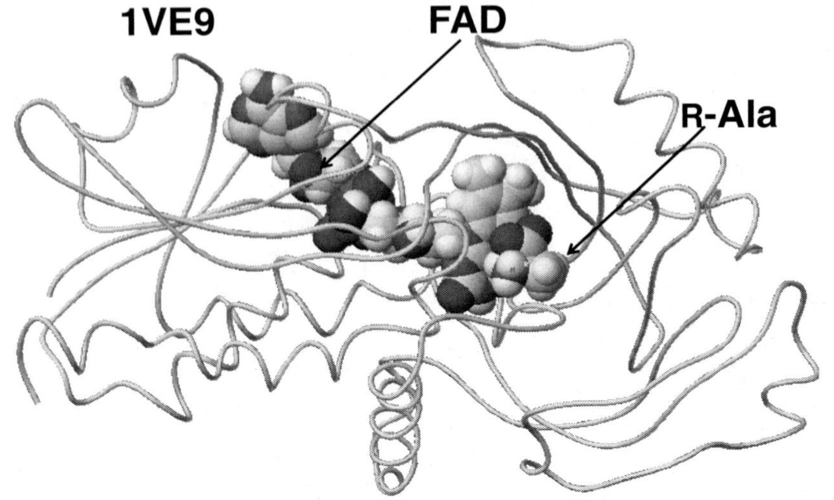

Figure 14. Optimized a portine kidney R-amino acid oxidase 1VE9 as a wire flame form with FAD and R-alanine using MM2 calculation. Symbols: see Figure 2. Atom size of FAD, and R-alanine molecules is shown as their full size.

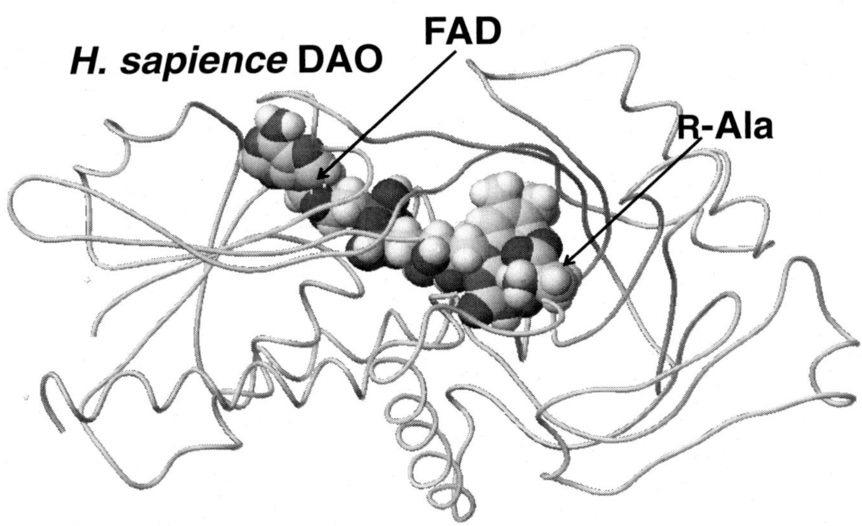

Figure 15. Optimized *H. sapience* R-amino acid oxidase as a wire flame form with FAD and R-alanine using MM2 calculation. Symbols: see Figure 2. Atom size of FAD, and R-alanine molecules is shown as their full size.

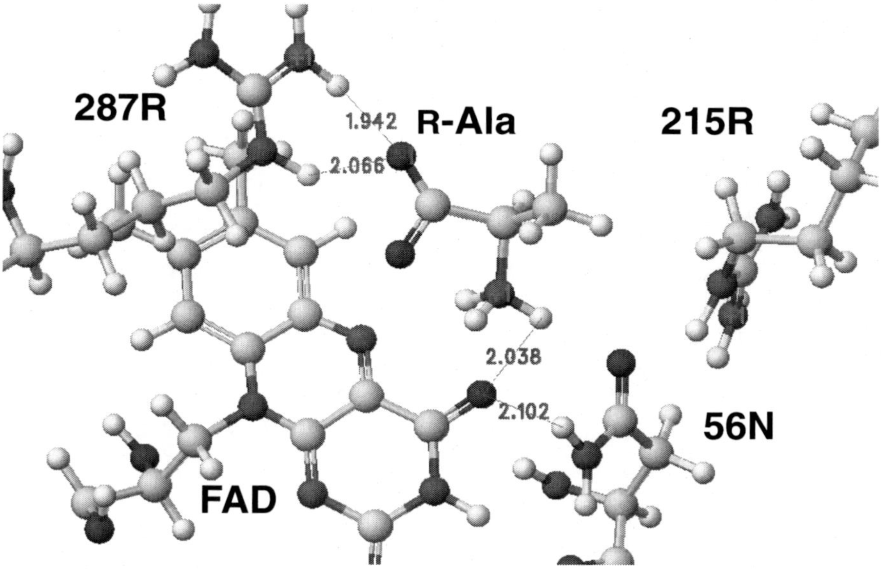

Figure 16. Extracted a R-alanine and 1C0P complex. Selected atomic distance is indicated. unit: Å.

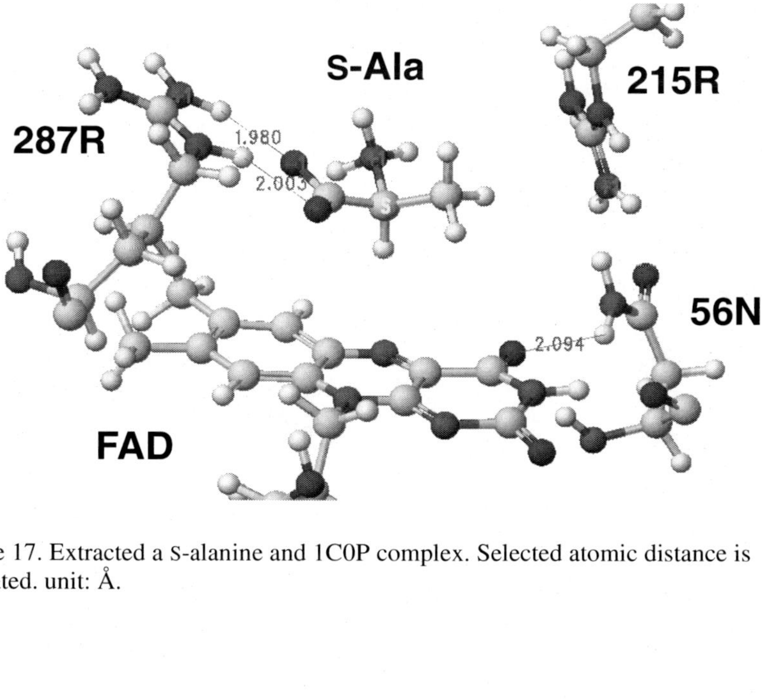

Figure 17. Extracted a S-alanine and 1C0P complex. Selected atomic distance is indicated. unit: Å.

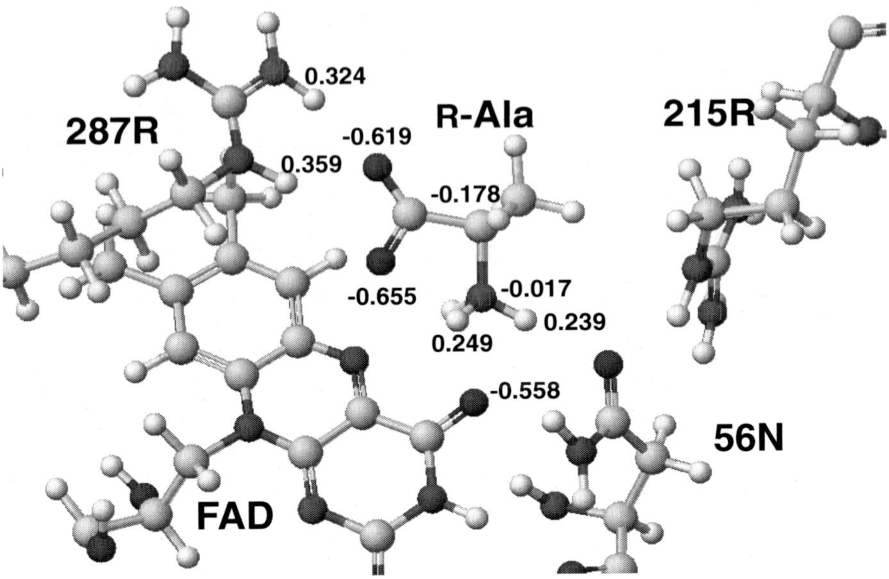

Figure 18. Atomic partial charge of selected atoms in Figure 16 calculated using MOPAC PM5 program. unit: atomic unit.

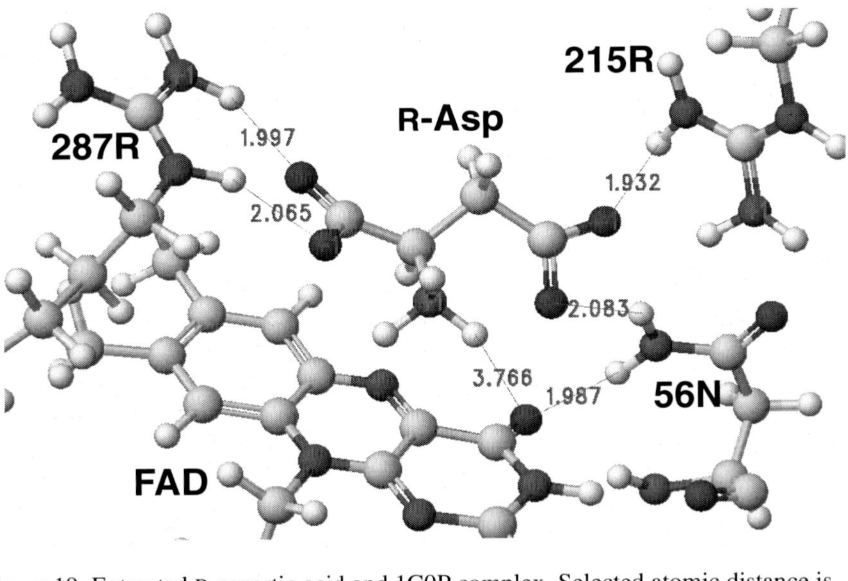

Figure 19. Extracted R-aspartic acid and 1C0P complex. Selected atomic distance is given. unit: Å.

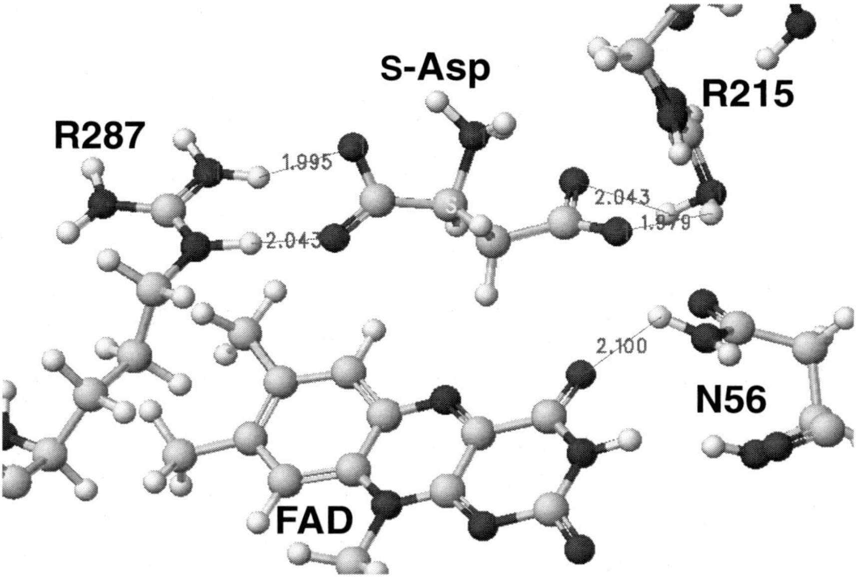

Figure 20. Extracted S-aspartic acid and 1C0P complex. Selected atomic distance is given. unit: Å.

The extracted structure of the S-amino acid M215R complex optimized using the MM2 calculation is shown in Figure 17. Similar to R-alanine, the carboxyl oxygen atoms in alanine are in tight contact with the guanidyl group hydrogen atoms of arginine. However, the amino group is located at the opposite-side of FAD, and consequently, no oxidation reaction can occur. The entire conformation was obtained after MM2 optimization after replacing R-alanine with S-alanine, superimposed at their carboxyl group.

Atomic partial charge on selected atoms, shown in Figure 12 was calculated, and the results are shown in Figure 18. Calculations were performed using the PM5 program of MOPAC after locking the extracted molecules. In case the molecules were not locked, the PM5 calculation re-optimized the structures, which resulted in incorrect information.

The structure was optimized after replacing R-alanine with R-aspartic acid. FAD was extracted and amino acid residues within 3Å from R-aspartic acid were locked, and the extracted structure was optimized to calculate the atomic partial charge. The extracted conformation is shown in Figure 19.

R-aspartic acid was replaced with S-aspartic acid and the entire structure was optimized. Extracted molecules are shown in Figure 20, and it is clear that the M215R enzyme is able to oxidize only R-amino acids. At every instance of amino acid replacement, the entire structure needs to be optimized, following which atomic distances can be measured and atomic partial charge (apc) can be calculated. This process allows the study of selective enzyme reactivity.

The feasibility of an enzyme reaction can be estimated from the change in the apc of the atoms at the reaction center. The correlation between Δapc of the α1 carbon or nitrogen of the amino group and log k_{cat}/K_m are 0.921 and 0.861 (n = 7), respectively [8].

Study of the Enzyme Reactivity of Alcohol Oxidoreductase 4W6Z and 1MG5

Alcohol dehydrogenase belongs to the oxidoreductase family, a class of enzymes responsible for the catalysis of biological oxidation and reductions. However, the reaction mechanisms were not clearly explained. Therefore, the stereo structures of a zinc-dependent *Saccharomyces cerevisiae alcohol dehydrogenase* (4W6Z) and a zinc-independent *Drosophila melanogaster dehydrogenase* (1MG5) were downloaded, and optimized using MM2 program. The structure was optimized after replacing original alcohol to another alcohol and the entire structure was optimized as done for DAO. At every instance of alcohol or aldehyde replacement, the entire structure needs to be optimized, following which atomic distances can be measured. The coenzyme and substrates within 3Å from the substrate were extracted, locked, and apc on these molecules were calculated using PM5.

The apc on the α-carbon of the alcohol demonstrated best agreement with the enzyme activity, V_{max}/K_m ($min^{-1}mg^{-1}$), where the correlation coefficients were found to be 0.861 and 0.838 (n = 9) for 4W6Z and 1MG5, respectively. Further, in the reduction process, the apc on the C5 and α-carbon atoms of the pyridine ring demonstrated a good agreement with the enzyme activity. The correlation coefficients for this process were 0.875 and 0.915 (n = 9) for 4W6Z and 1MG5, respectively [10].

Study of the Enzyme Reactivity of ELI3-2 Aromatic Alcohol $NADP^+$ Oxidoreductase

ELIS3-2 expressed in *Escherichia coli* is a biochemical function of *Arabidopsis thaliana* defense-related protein and an aromatic alcohol $NADP^+$ oxidoreductase. This enzyme demonstrated a strong preference for various aromatic aldehydes over corresponding alcohols [9].

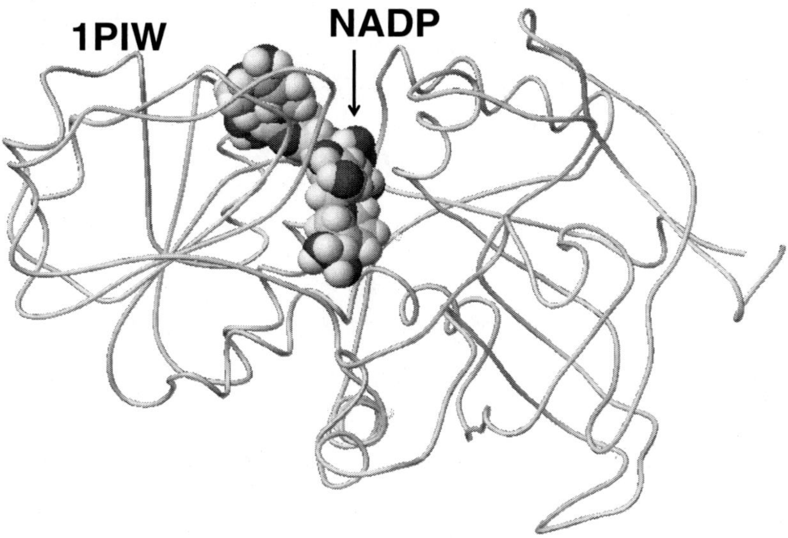

Figure 21. Structure of downloaded and optimized *cinnamyl alcohol dehydrogenase* 1PIW using MM2 calculation. Symbols: see Figure 2. 1PIW is shown as a wire flame form. Atom size of NADP is shown as their full size.

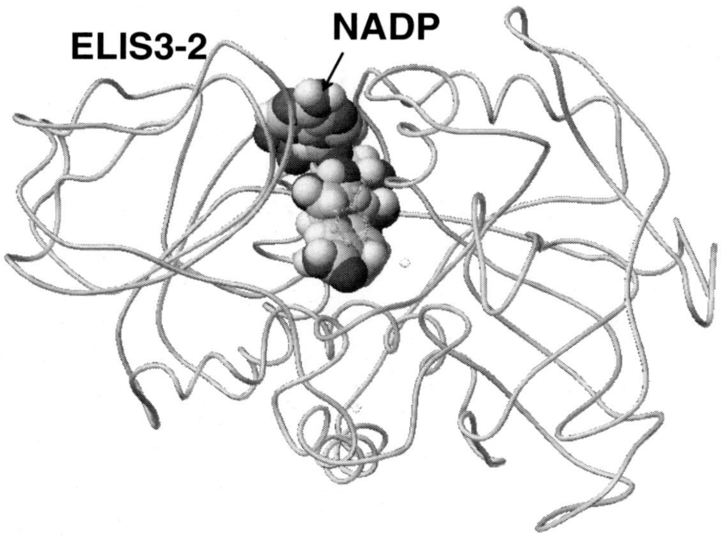

Figure 22. Constructed and optimized ELI3-2 structure using MM2 calculation. Symbols: see Figure 2. 1PIW is shown as a wire flame form. Atom size of NADP is shown as their full size.

Table 2. Comparison of sequences between chain A of porcine kidney D-amino acid oxidase 1VE9 and *H.sapience* D-amino acid sequence NP001908

Alignment statistics for match

Method	Identities	Positives	Gaps
Compositional matrix adjust.	294/347(85%)	321/347(92%)	0/347(0%)

Features:

```
NP001908     MRVVVIGAGVIGLSTALCIHERYHSVLQPLDIKVYADRFTPLTTTDVAAGLWQPYLSDPN  60
1VE9  1      MRVVVIGAGVIGLSTALCIHERYHSVLQPLDVKVYADRFTPFTTTDVAAGLWQPYTSEPS  60
              EEEEE  HHHHHHHHHHHHHHHH      EEEEEE  GGGGGGGG

NP001908 61  NPQEADWSQQTFDYLLSHVHSPNAENLGLFLISGYNLFHEAIPDPSWKDTVLGFRKLTPR 120
1VE9  61     NPQEANWNQQTFNYLLSHIGSPNAANMGLTPVSGYNLFREAVPDPYWKDMVLGFRKLTPR 120
              HHHHHHHHHHHHHHHH      GGGG EEEEEEEEE             EEEEE HHH

NP001908 121 ELDMFPDYGYGWFHTSLILEGKNYLQWLTERLTERGVKFFQRKVESFEEVAREGADVIVN 180
1VE9  121    ELDMFPDYRYGWFNTSLILEGRKYLQWLTERLTERGVKFFLRKVESFEEVARGGADVIIN 180
             HHH      EEEEEEEEEEHHHHHHHHHHHHHHH   EEEE    HHHHHHH    EEEE

NP001908 181 CTGVWAGALQRDPLLQPGRGQIMKVDAPWMKHFILTHDPERGIYNSPYIIPGTQTVTLGG 240
1VE9  181    CTGVWAGVLQPDPLLQPGRGQIIKVDAPWLKNFIITHDLERGIYNSPYIIPGLQAVTLGG 240
              GGGGGGGGG     EEEEEEEEEEE      EEEE           EEEE    EEEE

NP001908 241 IFQLGNWSELNNIQDHNTIWEGCCRLEPTLKNARIIGERTGFRPVRPQIRLEREQLRTGP 300
1VE9  241    TFQVGNWNEINNIQDHNTIWEGCCRLEPTLKDAKIVGEYTGFRPVRPQVRLEREQLRFGS 300
               EE        HHHHHHHHHHHHHHHHGGGGGEEEEEEEEEEEE    EEEE

NP001908 301 SNTEVIHNYGHGGYGLTIHWGCALEAAKLFGRILEEKKLSRMPPSHL  347
1VE9  301    SNTEVIHNYGHGGYGLTIHWGCALEVAKLFGKVLEERNLLTMPPSHL  347
                EEEEEE GGGHHHHHHHHHHHHHHHHHHHHHHHH

Second structure E: beta-parallel; H:right-handed alpha; G: 310 helix
```

The stereo structure of ELI3-2 (CADHB ARATH, Q02972) was constructed from a stereo structure of cinnamyl alcohol dehydrogenase 1PIW (ADH6 YEAST, Q04894) (Figure 21). The apo and holo structures of the NADP(H)-dependent cinnamyl alcohol dehydrogenase from *Saccharomyces cerevisiae* indicated an amino acid identity of 33.4% and shared a similarity of 122 amino acids [7]. The stereo structure thus (Figure 22) was similar to that of 1PIW. Atomic distance and atomic partial charge were related to the enzyme reactivity and showed a high correlation coefficient of 0.899 (n = 6) between V_{max}/K_m and the C4 hydrogen of the pyridine ring [10].

When the reaction cavity of the aliphatic alcohol dehydrogenase is smaller than that of an aromatic alcohol dehydrogenase, the original cavity size can be expanded by modification of the original substrate trifluoroacetic acid. For this purpose, three fluorine atoms are changed to methyl groups and the entire structure is optimized using the MM2 calculation. Furthermore, these methyl group hydrogen atoms can be converted to methyl groups, and the entire structure can be optimized before replacement with aromatic alcohols. This process allows expansion of the cavity size without interfering with the reaction center.

Aromatic aldehydes used here were 2-methoxybenzalehyde, 3-methoxybenzoladehyde, salicylaldehyde, benzaldehyde, cinnamaldehyde, and 4-coumaraldehyde. The apc of the aldehyde oxygen, coenzyme pyridine ring para-position carbon and hydrogen, and zinc demonstrate its involvement in the enzyme reaction. The further process was identical to that for M215R mutant mentioned earlier. Thus, given that sequence data are available, it is possible to construct a stereo structure of the human enzyme based on a similar enzyme of different spacies.

CONCLUSION

An updated version of the MM2 program enhanced hydrogen bonding interactions. This allows the construction of stereo structure of mutants from a known stereo structures and sequence data, which bear similar enzyme reaction capabilities. Optimization of stereo structure of the docked complex between the enzyme and its substrate by using MM2 calculations demonstrated the tightness of the complex. Further the calculations of atomic partial charge using MOPAC PM5 program permits quantitative analysis of enzyme reactivity.

REFERENCES

[1] Jayaraman, A.; Yarmush, M.L.; Roth, C.M. Molecular bioengineering *Ind. Eng. Chem. Res*. 2002, *41*, 441-455.

[2] Hanai, T. *Quantitative in silico chromatography, computational modeling of molecular interactions*, Royal Society of Chemistry, Cambridge, 2014, pp. 338.

[3] Sacchi, S.; Lorenzi, S.; Molla, G.; Pilone, M.S.; Rossetti, C.; Pollegioni, L. Engineering

[4] the substrate specificity of D-amino acid oxidase, *J. Biol. Chem*. 2002, *277*, 510-516.

[5] RCSB Protein Data Bank, www.rcsb.org/pdb/

[6] Hanai, T. Quantitative *in silico* analysis of molecular recognition and reactivity of D-amino acid oxidase, *Internet Electronic Journal of Molecular Design*, 2006, *5* (5) 247-259.

[7] Canepa, C.; Bach, R.D. A rate expression for enzymatic reactions. The dynamic relation between coupled oscillators at the active and the rate enhancement of enzyme-catalyses process. *Phys. Chem. Chem. Phys.* 2001, *3*, 4072-4081.

[8] NP 001908, D-Amino acid oxidase [Homo sapiens], www.ncbi. nlm.gov/entrez/query.fcgi?CMD.

[9] Hanai, T. Quantitative *in silico* analysis of enzyme reactions: comparison of D-amino acid oxidase and monoamine oxidase, *American Biotechnology Laboratory*, 2007, *25* (11) 8-13.

[10] Imre E. Somssich, Petra Werert, Siegrid Kiedroqwiski, Klaus Hahlbrock, Arabidopsis thaliana defence-related prtein ELIS-3 is an aromatic alcohol: NADP+ oxidoreductase, *Proc. Natl. Acad. Sci. USA*, 1996, *93*, 14199-14203.

[11] Hanai, T. Quantitative *in silico* analysis of alcohol dehydrogenase reactivity, *Journal of Drug Design and Research*,2014, *2*(1), 1006 (1-7).

In: Molecular Mechanics and Modeling
Editor: Priscilla Watkins
ISBN: 978-1-63483-388-2

Chapter 3

QUANTITATIVE *IN SILICO* ANALYSIS OF ALANINE RACEMASE REACTIVITY

Toshihiko Hanai*

Health Research Foundation,
Research Institute for production Development 4F,
Sakyo-ku, Kyoto, Japan

ABSTRACT

Downloaded structure of alanine racemase was optimized using a molecular mechanics calculation and used as an initial enzyme, then arginine-219 was replaced with other amino acids to construct the mutants. The atomic distances, bond angles, and apc of the mutants were correlated with enzyme reactivity and used to study the enzyme reaction mechanism. The alanine carboxyl group contacted lysine-39 and arginine-136. The alanine amino group contacted the oxygen molecules of pyridoxal 5'-phosphate aldehyde and phosphate, and aspartic acid-171. These amino acid residues contacted alanine with Coulombic force and may contribute to the enolation, rather than directly affecting the α-hydrogen exchange. The alanine α-hydrogen seldom contacted tyrosine-265'. The enzyme reactivity was quantitatively related to the bond angle of the substituted alanine. Although the apc can be used for quantitative analysis, the slope was small to predict the selective reactivity of the mutants. Which tyrosine is directly involved in the reaction seems to

* Email: thanai@kf7.so-net.ne.jp.

depend on the selected enzyme and mutant. A limited molecular dynamic calculation supported the results obtained using the molecular mechanics calculation.

Keywords: Alanine racemase, mutant, 116f, 115g, reactivity, *in silico*

INTRODUCTION

Quantitative analysis of enzyme reactions allows for the effective design of new drugs and the development of mutants for the preparation of practical immunoassays. The *in silico* analysis of molecular interactions in liquid chromatography provides quantitative methodology about the molecular recognition of proteins. Furthermore, quantitative analysis of changes in the atomic partial charge (apc) of key atoms used to study chemiluminescence intensity [1] can be used to study enzyme reactivity. Previous quantitative analysis of the selectivity of R-amino acid oxidase indicated that the selective reactivity correlated well with the apc of the substrate's key atoms [2]; the guanidino group of arginine is the positively charged site and the carboxyl group of an ionized amino acid is the negatively charged site. The substrates are pulled into the enzyme reaction center near the coenzyme flavine adenine dinucleotide (FAD) by Coulombic force. The carbonyl group of FAD is the electron donor and the ionized amino group of the substrate is deaminated at this site [3]. Deprotonation of the α-carbon has also been proposed to occur in the R-amino acid oxidase reaction process [4]. Deprotonation seems to be an important process in racemization. Pyridoxal 5'-phosphate (PLP) is a coenzyme of racemase. The reaction mechanisms of an R-amino acid oxidase and a racemase appear to be similar. Therefore, in the present study, I analyzed the reaction selectivity of alanine racemase mutants.

The racemization mechanism of amino acids was studied using arylglycines as the model amino acids. The zwitterion had the greatest reactivity. The inductive effect of the NH_3^+ group stabilizes the incipient carbanion, and reduces the electron density of the α-carbon. Removal of the proton attached to the α-carbon leaves behind a planar carbanion. The introduction of an electron-negative substituent on the α-carbon atom or on the amino group promotes racemization. Ionic strength is modestly affected by this reaction. The push-pull mechanism was proposed for racemization of arylglycines in sodium phosphate buffer [5]. The racemization of alanine by alanineracemase is thought to be proceeded by an initial transaldimination, followed by α-hydrogen (αH) abstraction to give the resonance-stabilized carbanion. Deprotonation on the opposite face yields the antipodal aldimine with subsequent release of the isomerized alanine moiety. An isotope effect accompanies abstraction of the α-deuteron, but no isotope effect is observed in the protonation step. Therefore, differential active sites for R- and S-alanine have been proposed. The detailed mechanistic features, however, remain unclear [6]. The crystal structures of alanine racemase bound with reaction intermediate analogs, *N*-(5'-phosphopyridoxyl)-S-alanine and *N*-(5'-phosphopyridoxyl)-R-alanine, were determined. Tyrosine-265' (Y265') and lysine-39 (K39) are the catalytic bases for abstraction of the α-proton from S- and R-alanine. The alanine carboxyl group directly contributes to mediate proton transfer between the two catalytic bases K39 and Y265'. The protonated carboxylic group is readily formed to stabilize the α-carbanion [7]. PLP enzymes exist in their resting state as Schiff bases with an active site lysine residue. The incoming amine-containing substrate displaces the lysine ε-amino group from the internal aldimine. Y265' and K39 are likely to be the two acid/base catalysts in a two-base mechanism of racemization [8]. This means that the amino group of the substrate amino acid should be close to the binding site. The efficient catalysis of deprotonation of S-alanine by PLP and by alanine racemase was studied using a computational chemical method. Large solvation effects of the external aldimine cofactor lower the p*K*a of the α-amino acid. Unprotonated PLP increases the acidity of the α-amino acid [9]. Co-enzyme PLP contributes to the racemization reaction process. The electron delocalization in PLP is affected by R136 and R219 [10].

Racemization occurs in vitro without an enzyme [5, 11]. If the reaction process follows the same scheme in vivo, the ionized alanine should be located near the aldehyde of PLP or the lysine (K39) nitrogen binds with PLP and replaces K39. Racemization occurs when a substrate alanine must compete with K39 to bind with the PLP aldehyde. An αH on the substrate alanine indicates the possibility of deprotonation inside the alanine racemase. Here, the reaction selectivity of alanine racemase mutants was quantitatively studied based on the above-reported results and the selectivity of the mutants was compared according to the stereo structure of alanine racemase. The selective reactivity was determined from data obtained using the mutants, arginine-219-lysine (R219K), arginine-219-alanine (R219A), and arginine-219-glutamic acid (R219E), to investigate the role of R219 in catalysis [12]. Different conformations, including K39-conjugated PLP were constructed to make complexes with either S- or R-alanine, and the structures were optimized. The conformation of alanine was then examined to study the reactivity of the mutants. To study the racemization mechanism, we examined whether PLP bound with K39 prior to exchange with the alanine or whether PLP supports electron transfer or localization for smooth racemization. The apc of key atoms was used to analyze the electron shift. A molecular dynamics program was used to observe the possible competition between alanine and K39 to bind with the PLP aldehyde.

EXPERIMENTAL

The structures of alanine racemase were downloaded from the RCSB Protein Data Bank [13], and any irregular residues and atoms were corrected. The PLP-alanine complex was then separated into individual molecules. Mutants were constructed by replacing R219 with different amino acids. The stereo structure of the complexes was optimized using molecular mechanics force-field parameters (MM2) of the CAChe program (Fujitsu, Tokyo, Japan). The minimum energy level was 10^{-7} kcal/mol. Furthermore, the apc of key atoms was calculated using MOPAC PM5 of the CAChe program after extraction of the alanine, PLP, and surrounding amino acid residues within 3Å of the alanine. The possible movement of alanine for racemization was studied using the molecular dynamics program of CAChe program. A Dell Optiplex GX270 computer (Dell, Japan) with a 2.4 MHz CPU was used.

Results and Discussion

Seven stereo structures of alanine rascemase found in the PDB archive, 1l6f, 1n1u, 1ftx, 1sft, 2dy3, 1l6g, and 2sfp were studied. Only 1sft seemed to comprise the complete structure. A yeast alanine racemase structure, 1l6f, was selected, however, even though K129 was not fixed correctly due to the complex form with S-alanyl-pyridoxal-5'-phosphase. Another candidate, 1l6g, included R-alanine conjugated with PLP, but the PLP ring was not aromatic.

First, water molecules were eliminated from the crystal structure because no water molecules exist near the reaction site and charge-transfer directly occurs between two molecules, and the substituted alanine was freed from the PLP complex, and the fixation of K129 was corrected. The cysteine sulfur was fixed. The mutants were constructed by replacing R219 with alanine, glutamic acid, or lysine, and the structures were optimized using the MM2 force field of the CAChe program. The 1l6g PLP aliphatic ring was converted to an aromatic ring, and the stereo structures were optimized using MM2. The alanine racemase was crystallized as double molecules (dimers), and not single molecules (monomers). The reaction center, therefore, existed at two sites, as shown in the corrected stereo structure of 1l6f (Figure 1). The first site is related to the first PLP in the sequence table, and the second site is related to the second PLP in the sequence table. At the first site, K39, Y43, R136, D171, Y354, Y265', Y284', and M312' were located within 3Å of the S-alanine. At the second site, K39, Y43, R136, Y354, Y265', C311', and M312' were located within 3Å of the S-alanine. The extracted amino acid residues, PLP, and S-alanine at the first and second sites are shown in Figures 2 and 3. These molecules were used to calculate the apc using MOPAC PM5. Substrate 219, behind the PLP, was not included as a substrate within 3Å of the alanine. The reaction process was studied at both sites separately.

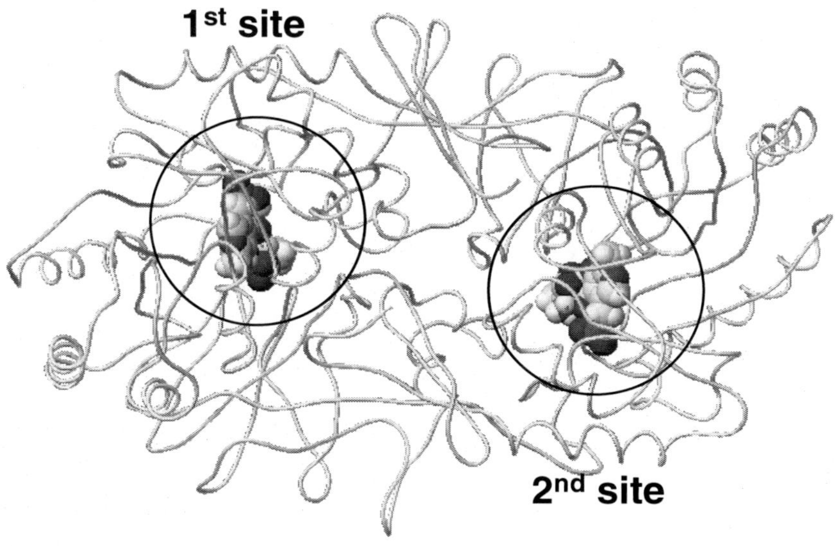

Figure 1. Reaction sites of alanine racemase dimmer. 1st site is 1st PLP location and 2nd site is 2nd PLP location. Light gray ball: carbon, small white ball: hydrogen, dark gray ball: nitrogen, black ball: oxygen, large black ball: phosphate. Proteins are indicated as wire flame form.

The enzyme reactivity (kinetic parameter), log k_{cat}/K_M (M^{-1} s^{-1}) from reference 12, was used to study the selective reactivity of the mutants, and was related to atomic distance, atomic bond angle, and the apc of the alanine of the different complexes. The reactivity values in the change from S-alanine to R-alanine were: wild-type 5.672 > R219K 4.977 > R219A 3.568 > R219E 2.886, and those in the change from R-alanine to S-alanine were wild-type 5.602 > R219K 4.978 > R219A 3.505 > R219E 2.778. The equilibrium constant $K = (k_{cat}/K_M)_{R\rightarrow S}/(k_{cat}/K_M)_{S\rightarrow R}$ was equal to 1. The correlation coefficient was 1.000 (n = 4) [12].

The stereo structure of the mutants indicated that the 219th amino acid residue was not directly involved in the racemase reaction because it locates behind PLP. The 219th amino acid, however, did seem to affect the electron localization of PLP.

After optimizing the complex structures, the atomic distance, bond angles around the α-carbon (αC) of the alanine, and the apc of selected atoms were used to evaluate factors involved in the selectivity of the racemase reactions of the mutants. The K39 amino group must compete with the amino group of the substrate amino acid, alanine. The atomic distances from the PLP aldehyde to the ε-amino group of K39 or the amino group of alanine were used to analyze the replacement reaction.

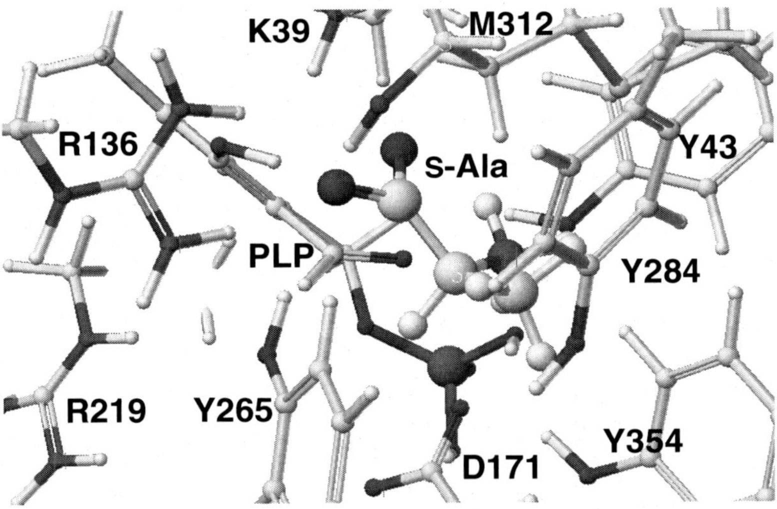

Figure 2. Amino acid residues of 1st site located within 3Å from s-alanine and R219. Gray ball: carbon, white ball: hydrogen, black ball: oxygen, dark gray ball: nitrogen, large black ball: phosphate, gray ball: sulfur; Ball size of alanine and phosphate is 4 times from residues.

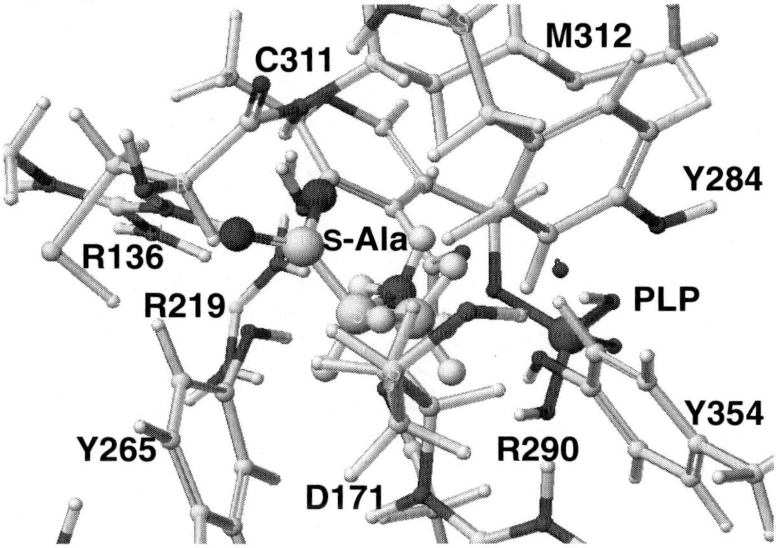

Figure 3. Amino acid residues of 2nd site located within 3Å of s-alanine and R219. Symbols: see Figure 2.

The possible electron transfer contact was studied from the optimized structures. At the first site of the wild-type alanine racemase, the S-alanine amino group contacted the oxygens of the PLP aldehyde and phosphate, the D171 carboxyl group, and the Y43 hydroxyl group. The alanine carboxyl group contacted the hydrogens of the K39 amino group, the Y265' and Y43 hydroxyl groups, the PLP hydroxyl group, and the R136 guanidino group. The K39 ε-amino group contacted the PLP hydroxyl group, but not, however, the PLP aldehyde. The αH contacted both oxygens of the PLP aldehyde and the Y265' hydroxyl group with a distance of approximately 2.4Å. At the second site of the wild-type alanine racemase, the S-alanine amino group contacted the oxygens of the PLP aldehyde and phosphate groups, as well as the Y354 and Y265' hydroxyl groups. The alanine carboxyl group contacted the hydrogens of the R136 guanidino group and the PLP hydroxyl group. The nearest neighbor of the αH was the oxygen of the Y354 hydroxyl group (2.905Å). The existence of different conformations at the two sites indicates selective racemization; that is, only one location may be practical for alanine racemization. The conformation of the mutants R219A, R219E, and R219K was similar to that of wild-type alanine racemase.

Table 1. Properties of alanine in complex with alanine racemase dimer

Complex	distance Å	apc au					bond angle degree		
	αH-αC	αH	αC	O*	O-	N+	BA1	BA2	BA3
S-Alanine	1.115	0.189	-0.202	-0.634	-0.579	0.007	99.21	114.49	108.64
R-Alanine	1.116	0.186	-0.240	-0.495	-0.664	0.026	91.64	117.63	108.83
mPLP, mY, site 1/1l6f									
R219A, sAl	1.110	0.225	-0.073	-0.677	-0.677	-0.038	105.42	110.97	110.65
R219E, sAl	1.109	0.226	-0.081	-0.668	-0.685	-0.038	105.57	110.75	111.38
R219K, sAl	1.110	0.212	-0.079	-0.688	-0.656	-0.048	105.11	112.01	111.37
WT, sAl	1.110	0.217	-0.078	-0.692	-0.658	-0.046	105.02	111.85	111.31
mPLP, mY, site 2/1l6f									
R219A, sAl	1.114	0.225	-0.142	-0.557	-0.714	-0.010	98.90	114.33	105.75
R219E, sAl	1.114	0.223	-0.148	-0.556	-0.715	-0.008	98.43	115.36	105.05
R219K, sAl	1.114	0.223	-0.143	-0.561	-0.709	-0.012	99.39	113.32	105.77
WT, sAl	1.114	0.224	-0.143	-0.561	-0.711	-0.012	99.21	113.57	106.01
iPLP, mY, site 1/1l6f									
R219A, sAl	1.112	0.219	-0.105	-0.624	-0.712	-0.036	104.64	111.94	111.00
R219E, sAl	1.112	0.197	-0.126	-0.646	-0.689	-0.050	104.37	112.52	111.74
R219K, sAl	1.112	0.195	-0.129	-0.643	-0.702	-0.052	104.65	113.62	111.60
WT, sAl	1.111	0.218	-0.106	-0.624	-0.712	-0.037	104.58	112.09	111.10
iPLP, mY, site 2/1l6f									
R219A, sAl	1.114	0.223	-0.139	-0.556	-0.699	-0.027	99.21	113.61	104.04
R219E, sAl	1.114	0.222	-0.140	-0.558	-0.700	-0.025	98.97	113.25	104.47
R219K, sAl	1.114	0.202	-0.120	-0.598	-0.681	-0.049	103.62	109.73	106.87
WT, sAl	1.114	0.208	-0.131	-0.570	-0.678	-0.034	102.12	111.10	106.88
mPLP, mY, site 1/1l6g									
R219A, RAl	1.114	0.200	-0.175	-0.552	-0.725	-0.024	97.93	113.40	112.26
R219E, RAl	1.111	0.203	-0.177	-0.551	-0.731	-0.017	97.61	113.93	112.84
R219K, RAl	1.114	0.199	-0.176	-0.551	-0.725	-0.024	97.89	113.18	112.39
WT, RAl	1.113	0.198	-0.176	-0.552	-0.725	-0.023	97.99	112.99	112.72
mPLP, mY, site 2/1l6g									
R219A, RAl	1.114	0.165	-0.135	-0.591	-0.731	-0.030	102.01	114.34	113.65
R219E, RAl	1.110	0.168	-0.141	-0.591	-0.730	-0.043	100.86	120.03	113.80
R219K, RAl	1.114	0.164	-0.134	-0.589	-0.731	-0.032	102.00	114.41	113.59
WT, RAl	1.114	0.164	-0.131	-0.590	-0.730	-0.033	102.20	113.99	113.61

O*: carbonyl oxygen; BA1: COO^--αC-NH_3^+; BA2: COO^--αC-Me; BA3: Me-αC-NH_3^+; R-& S-Alanine : free form ionized alanine; mPLP: molecular form PLP, mY: molecular form tyrosine; iPLP: ionized PLP.

From the above observations, it was not clear whether the contribution of Y265' is a direct push or pull on the αH. Which atom contributes to the push or pull of αH? At the first site, the PLP aldehyde may be involved in the push or pull of αH, and binding with the alanine amino group. At the second site, the hydroxyl group of Y354 may contribute to the reaction more than Y265'. Tyrosine-265' may contribute to proton transfer through the carboxyl group of alanine at the first site and through the amino group of alanine at the second site. The guanidino group of R139 and the aldehyde of PLP are likely more important than the K39 amino group and the Y265' hydroxyl group. The contribution of Y265' was also not clear for alanine racemase 2sfp (detail is not given).

If the direct push-pull of αH is critical, then the atomic distance between αH and αC should demonstrate the selective reactivity of the mutants. The atomic distance is given in Table 1 with other properties. The atomic distance between αH and αC at the two sites changed little across the mutants, and was not related to the reactivity. The aldimine form is considered to be the transition state. The alanine molecule becomes flat, and this phenomenon can be observed as a bond angle change around the αC. The original optimized bond angle of S-alanine was 322.34°. The bond angle was less than the original at the second site in all forms except the wild-type form, and it was a little wider at the first site. No meaningful correlation between the reactivity and total bond angle at the first and second sites was determined. The total bond angle was narrower at the first site, whereas the reactivity was higher. The alanine carboxyl group at both the first and second sites was correlated to the reactivity. The correlation coefficients for the bond angles of COO^--αC-NH^+_3 at the first and second sites were 0.997 (n = 4) and 0.890 (n = 4), respectively. Those for the bond angle COO^--αC-CH_3 at the first and second sites were 0.943 (n = 4) and 0.913 (n = 4), respectively. The bond angle change can be used to study mutant reactivity. Even if the reactivity in the change from R- to S-alanine was used, the behavior was the same. The first and second sites demonstrated different behaviors in the racemization of S-alanine. The correlation coefficients for the bond angles of COO^--αC-NH^+_3 at the first and second sites were 0.998 (n = 4) and 0.902 (n = 4), respectively. Those for the bond angles of COO^--αC-CH_3 at the first and second sites were 0.943 (n = 4) and 0.912 (n = 4), respectively.

Further studies were performed using the apc of key atoms calculated using PM5 of the MOPAC program. Meaningful correlation coefficients and the reactivity of the mutants obtained for the alanine carboxyl group and nitrogen, and the correlation coefficients between $(k_{cat}/K_M)_{S \to R}$ and carbonyl oxygen, ionized carboxyl oxygen, and ionized nitrogen at the first site were 0.990, 0.961, and 0.904 (n = 4), respectively. Those at second site were 0.973, 0.863, and 0.953, respectively. The correlation coefficients between $(k_{cat}/K_M)_{R \to S}$ and carbonyl oxygen, ionized carboxyl oxygen, and ionized nitrogen at the first and second sites were 0.996, 0.968, 0.912, 0.962, 0.934, and 0.960 (n = 4), respectively.

These correlation coefficients, however, did not indicate the selectivity of the two sites from the change in S-alanine to R-alanine or in the change from R-alanine to S-alanine based on the conformation of alanine. The reactivity was studied using the apc of the atoms of related residue. At the first site, the apc values of the oxygens of Y265', the PLP aldehyde, and D171 did not support the selective reactivity. At the second site, however, the apc values of the oxygens of the PLP aldehyde and the Y354 hydroxyl group indicated that the reaction could involve the transport of protons through the oxygens located close to the alanine amino group. The correlation coefficients were 0.928 (PLP aldehyde, n = 4) and 0.789 (Y354; n = 4). The Y265' and Y43 oxygens were not correlated. The correlation coefficients improved a little (r's = 0.938 and 0.806, respectively) in the change from R-alanine to S-alanine.

In the above conformations, αH contacted the PLP aldehyde oxygen at the first site, and the closest oxygen was the Y354 hydroxyl group at the second site. The contribution of Y265' was not clear from the results. The PLP ring nitrogen is involved in an intermolecular OHN hydrogen bond with an aspartic acid (Q222) side-chain and the phosphate group forms three hydrogen bonds with other amino acid residues of aspartate aminotransferase from *Escherichia coli* [14]. Such an acidic amino acid residue does not exist in alanine racemase 116f. The ionization effect of the PLP nitrogen and hydroxyl groups was studied for the quantitative analysis of the reaction selectivity. The conformation changed a little. At the first site of the wild-type alanine racemase, the S-alanine amino group contacted the oxygens of the PLP aldehyde and phosphate groups, the D171 carboxyl group, and the Y43 hydroxyl group. The alanine carboxyl group contacted the hydrogen of the K39 ε-amino group, and the Y265'and Y43 hydroxyl groups. The K39 amino group contacted the PLP ionized hydroxyl group, but not the PLP aldehyde. αH contacted the oxygen of the PLP aldehyde with a distance of 2.5Å. At the second site of the wild-type alanine racemase, the S-alanine amino group contacted the oxygens of the PLP aldehyde and phosphate, and the Y354 and Y265' hydroxyl groups. The alanine carboxyl group contacted the hydrogens of the R136 guanidino and the Q314 amino groups, and the oxygen of the PLP aldehyde. The nearest neighbor of αH was the Y354 hydroxyl group oxygen, with a distance of 2.677Å. The existence of different conformations at the two sites indicates selective racemization or that the different conformations at only one site are functional for alanine racemization. The conformation of mutants R219A, R219E, and R219K, was similar to that of wild-type alanine racemase. The correlation coefficients related to molecular shape were not as strong as those obtained using the molecular form of PLP. There was no meaningful correlation coefficient at the first site. The correlation coefficient for the COO^--αC-NH^+_3 bond angle was 0.863 (n = 4) at the second site, and that for the COO^--αC-CH_3 bond angle was 0.816 (n = 4) at the second site. The correlation coefficient of apc obtained for the ionized carboxyl group oxygen and that of the amino group hydrogen of the alanine was acceptable at only the second site (oxygen: r = 0.975 [n = 4]; hydrogen: r = 0.908 [n = 4]). Similar results were obtained for the reactivity in the change from R-alanine to S-alanine (oxygen: r = 0.976 [n = 4]; hydrogen: r = 0.909 [n = 4]). The result suggested that the pyridoxal ring has an ionizing effect at the second site.

The extracted complex conformations were further optimized using the MOPAC PM5 program after locking only the background atoms, the peptide connection atoms. In the wild-type complex at first site, the alanine carboxyl oxygen bound with K39 hydrogen, and alanine amino hydrogen bound with D171 ionized carboxyl oxygen. The atomic distances were 1.259 and 1.337 Å, respectively. The distance between the K39 ε-amino hydrogen and PLP aldehyde oxygen was 4.014 Å. The optimized conformation indicated that the aldimine is formed by enolation through K39 and D171, and Y265' does not contribute to the reaction. Similar results were obtained for the mutants.

At the wild-type second site, the expected conformation was not obtained. Alanine carbonyl oxygen contacted the PLP hydroxyl group, and the amino hydrogen located near the Y43 and Y354 hydroxyl groups. The PLP aldehyde, K39 ε-amino group, D171 carboxyl group, and Y265 hydroxyl group were far from S-alanine. It is therefore difficult to determine the reaction bases on the conformation alone.

Further studies were performed using the corrected structure of 1l6g. At the first site of wild-type alanine racemase, the S-alanine amino group contacted the oxygens of the PLP aldehyde, hydroxyl, and phosphate groups, and the oxygens of the D313 carboxyl group and the Y265' hydroxyl group. The alanine carboxyl group contacted the hydrogens of the Y284' and Y265'carboxyl groups. The K39 ε-amino group contacted the PLP hydroxyl group and the PLP aldehyde oxygen. The αH was close to the oxygen of the PLP aldehyde with a distance of approximately 3.1Å. At the second site of wild-type alanine racemase, the S-alanine amino group contacted the oxygens and hydroxyl group of the PLP aldehyde, as well as the oxygens of the D313 carboxyl and Q314 carbonyl groups. The alanine carboxyl group contacted the hydrogens of the R136 guanidino group, the Y265' and Y354 hydroxyl groups, and the Q314 amino group. The K39 ε-amino group contacted the PLP aldehyde oxygen, PLP hydroxyl oxygen, and D313 carboxyl group oxygen. The hydrogen acceptor nearest to the αH was the Q314 oxygen, with a distance of 3.182Å. The conformation of the mutants R219A, R219E, and R219K, was similar to that of wild-type alanine racemase. The COO^{-}-αC-CH_3 bond angle correlated with the reactivity of the mutants at the first and second sites of 116g (correlation coefficients = 0.942 and 0.843 [n = 4], respectively). The αH did not make contact with any oxygens, but the apc correlated well with the reactivity (r's = 0.928 and 0.866 [n = 4] at the first and second sites, respectively). A similar correlation was obtained in the reactivity for the change from R-alanine to S-alanine.

Each monomer has a reaction site and the two reaction sites have different conformations. The conclusion about the reaction mechanism depended on the reaction site selected for the computation. Based on the literature, Y265' is considered a key residue, but the above results did not confirm its importance. Rather the PLP aldehyde, the hydroxyl group of another tyrosine residue, and the R136 guanidino group seemed to be directly involved in the reaction. If the contribution of Y265' is negligible, a single protein can work as a racemase. One monomer of the optimized 116f mutants was selected, and the reactivity was studied using the atomic distance, bond angles, and apc of selected atoms. The details of the complex conformations are explained using R219K as the model due to its high reactivity.

In the optimized structure of a mutant, R219K without alanine, where PLP and tyrosine were in their molecular form (mPLP, mY), the atomic distance between the K39 ε-amino group and the PLP aldehyde was 3.7Å. The guanidino group of R136 contacted the oxygens of the PLP hydroxyl and aldehyde groups with an atomic distance of 2Å. The K39 ε-amino group contacted the Y43 and Y354 hydroxyl groups. In the optimized R219K with S-alanine (Figure 4), the alanine amino group contacted the PLP aldehyde, Y43 hydroxyl group, and the D171 carboxyl group with an atomic distance of 2Å. The atomic distance between the K39 ε-amino group and the alanine carboxyl group was 2Å. The R136 guanidino group contacted the PLP hydroxyl group, and the alanine carboxyl group. For S-alanine, the hydrogen acceptor group nearest to αH was the PLP aldehyde, and the nearest hydrogen donor group was the R136 guanidino or K39 ε-amino groups, with an atomic distance of approximately 2Å. The amino group of R-alanine was rejected by the R136 guanidino group, and contacted the Y43 hydroxyl oxygen.

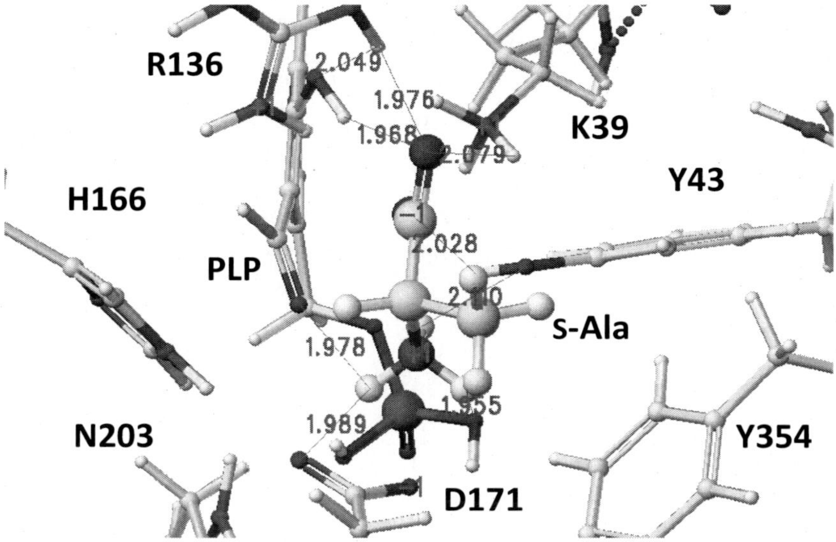

Figure 4. Conformation of 1st site with s-alanine. Atomic distance: Å. Symbols: see Figure 2.

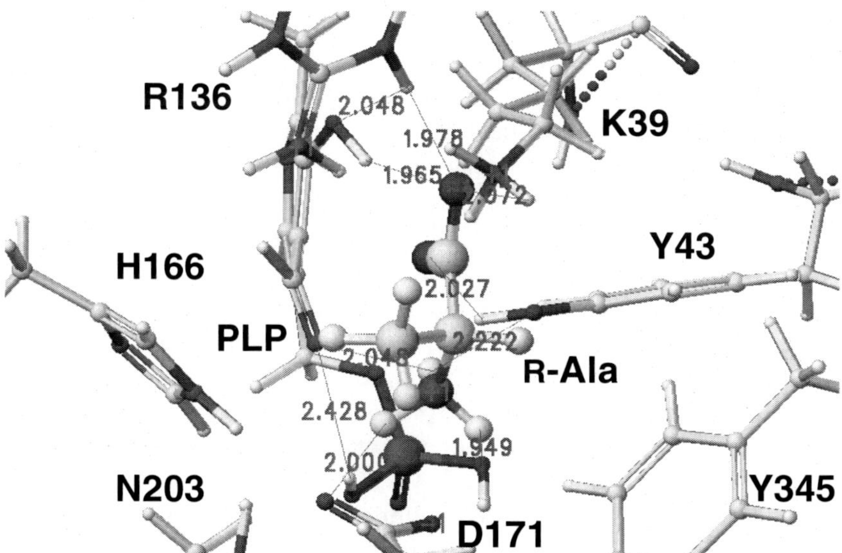

Figure 5. Conformation of 2nd site with R-alanine. Atomic distance: Å, Symbols: see Figure 2.

The carboxyl group of R-alanine contacted the K39 ε-amino group, the R136 guanidino group, and the Y43 hydroxyl group. The R136 guanidino group contacted the PLP hydroxyl group (Figure 5). For R-alanine, the hydrogen acceptor group nearest the αH was the Y43 hydroxyl group, and the nearest hydrogen donor group was the R136 guanidino group, with atomic distances of 3.3Å and 2Å, respectively. The alanine methyl group was above the reaction center and did not block the amino group from contacting key oxygens. The conformation of the optimized structure of R219K with ionized PLP, was similar to that of R219K optimized using the molecular form of PLP.

Based on the literature, Y265' may be ionized by a solvent effect under physiologic conditions even if the p*K*a value of tyrosine is 10.07. If Y265' can be ionized, then nearby Y43 and Y354 may also be ionized. The complexes with the mPLP and ionized Y43 and Y354 (iY) were optimized, and the conformations were analyzed. There was little difference in the conformation compared to the above two complexes. The distance of Y43 from the alanine was more than 3Å from the αH. The nearest hydrogen acceptor was the D171 oxygen, with a distance of 2.5Å. The nearest hydrogen donor group was the R136 guanidino and K39 ε-amino groups, with a distance of 2Å. The K39 ε-amino group was located far from the PLP aldehyde. The oxygens of D171, Y43, and Y354 contacted the alanine amino group. The alanine carboxyl group contacted the K39 ε-amino and R136 guanidino hydrogens. The PLP aldehyde oxygen was close to the alanine amino group. The conformations of the R-alanine complexes were similar to those of the S-alanine complexes. The distance between αH and Y43 oxygen was 2.8Å. The R-alanine amino group contacted the PLP aldehyde, D171 carboxyl, and Y43 hydroxyl oxygens. The R-alanine carboxyl group contacted the R136 guanidino and K39 ε-amino groups, with an atomic distance of 2Å. If atomic distance indicates the reactivity, the D171 carboxyl, and R136 guanidino groups are directly involved in the reaction. The PLP aldehyde may be directly involved in the reaction, but the K39 ionized ε-amino group does not compete with the alanine amino group. The conformations of the other mutants and wild-type were similar to that of R219K.

When PLP, Y43, and Y354 were all ionized (iPLP, iY), the conformation became tighter. The alanine amino group contacted the oxygens of Y43, D171, and the PLP aldehyde. The alanine carboxyl group contacted the K39 and R136 amino hydrogens. For S-alanine, the nearest hydrogen acceptor was the D171 oxygen, and the nearest hydrogen donor was the R136 guanidino and

K39 ε-amino groups with an atomic distance of 2Å. The atoms contacting the alanine amino and carboxyl groups for R-alanine were similar to those of S-alanine. The αH contacted the Y354 oxygen with an atomic distance of 2.6Å. The nearest hydrogen donor was the R139 guanidino and K39 ε-amino groups. The racemase reaction may have occurred by either direct pull-and-push or by electron transfer through the amino and carboxyl groups.

The complex conformation using PLP-bound K39 was very different from the above results. The PLP phosphate shifted toward R136. The alanine amino group contacted the PLP and Y354 hydroxyl groups, but not the Y43 hydroxyl group which was more than 3Å away from the alanine. The alanine carboxyl group contacted the PLP phosphate and the R136 guanidino group. The hydrogen donor nearest αH was the R136 guanidino group, and the nearest hydrogen acceptor was the PLP oxygen. The distance between the αH and these groups was 3.7Å. There was no hydrogen acceptor group nearby. If the bound K39 can be replaced with an amino acid, albumin glycation can be protected and glycated albumin can be cleaved, but this has not been done. The D171 carboxyl group was located far from the R136 guanidine group (Figure 6). The conformation indicated that replacing alanine with K39 would be difficult. The conformations of other mutants and the wild-type were similar to that of R219K. The reaction selectivity could not be quantitatively analyzed from the conformation. If direct deprotonation of αC is critical for the pull-and-push mechanism, then the atomic distance between αC and αH may indicate the reactivity of the mutants. The atomic distance changed little within the mutants after the complex conformations were optimized using either the ionized or molecular forms of PLP or with the K39-PLP complex, even if some conformations had a short atomic distance. Similar results were obtained for both S- and R-alanines.

When the racemase activity was high, the alanine structure had to flatten against the imine structure. The bond angle around the αC of alanine was measured after optimization of these complexes. A significant change in the bond angle was observed between the carbons of COO^- - αC – NH_3^+. The original COO^- - αC - NH_3^+ bond angles of ionized S- and R-alanine were 99.21 and 91.64 °, respectively, as summarized in Table 2. The ionization effect was studied by the correlation between the properties and reactivity $(k_{cat}/K_M)_{S\rightarrow R}$ and $(k_{cat}/K_M)_{R\rightarrow S}$. The correlation coefficients between COO^--αC-NH_3^+ and $(k_{cat}/K_M)_{S\rightarrow R}$ with mPLP and mY, mPLP and iY, iPLP and mY, iPLP and iY, and the K39-PLP complex were 0.895, 0.836, 0.772, 0.943, and 0.809 (n's = 4), respectively.

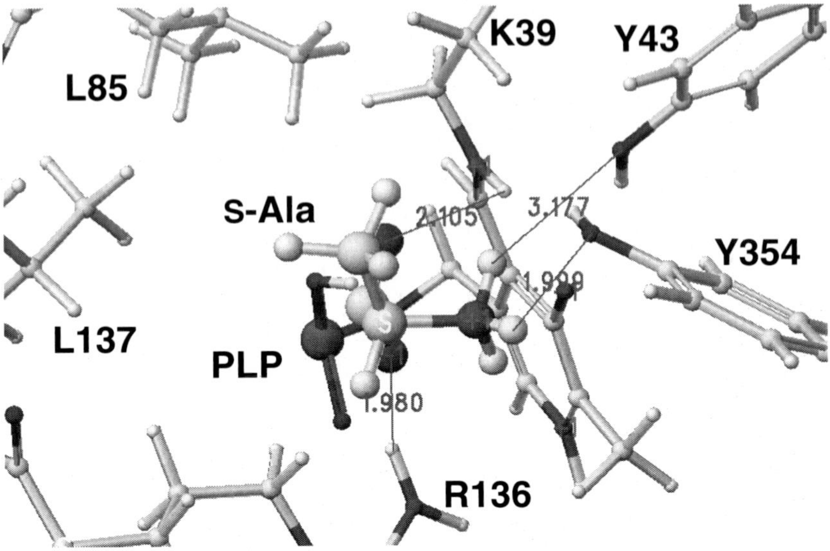

Figure 6. Conformation of R219K with PLP binds K39 lysine. Atomic distance: Å. Symbols: see Figure 2.

The correlation coefficients with $(k_{cat}/K_M)_{R\rightarrow S}$ were 0.906, 0.851, 0.754, 0.949, and 0.806, respectively. When R-alanine was used as a part of the complexes, the correlation coefficients for the angle with $(k_{cat}/K_M)_{S\rightarrow R}$ were 0.902, 0.967, and 0.705 for the mPLP and mY, iPLP and mY, and iPLP and iY complexes, respectively. The values increased with $(k_{cat}/K_M)_{R\rightarrow S}$, and the values were 0.909, 0.971, and 0.720, respectively.

The contribution of apc depended on the initial structure extracted. When a part of monomer was used, several apcs clearly demonstrated the contribution based on the correlation coefficients between the apc and the reactivity. The reactivity values in the change from S-alanine to R-alanine correlated well with the apc of the ionic amino nitrogen and αC of S-alanine optimized with the molecular form of PLP and tyrosine (r's = 0.914 and 0.789 [n = 4], respectively). The correlation coefficient increased to 0.975 and 0.953 for the R-alanine carboxyl oxygen and αC, respectively. The correlation coefficients increased when the enzyme reactivity values in the change from R-alanine to S-alanine were analyzed (r's = 0.921, 0.806, 0.971, and 0.960, respectively). When the tyrosines were ionized, the correlation coefficient with reactivity decreased (r's = 0.863 and 0.756 for S-alanine [S→R], and 0.876 and 0.772 for s-alanine [R→S]).

Table 2. Properties of alanine in complex with alanine racemase monomer 1l6f

Complex	distance Å		apc (au)				bond angle degree		
	αH-αC	αH	αC	O*	O-	N+	BA1	BA2	BA3
mPLP,mY									
R219A+SAl	1.115	0.203	-0.071	-0.628	-0.713	-0.063	103.60	111.55	109.87
R219E+SAl	1.116	0.181	-0.067	-0.718	-0.616	-0.023	101.72	110.86	109.72
R219K+SAl	1.115	0.198	-0.069	-0.632	-0.712	-0.063	103.63	112.14	110.05
WT+SAl	1.115	0.198	-0.073	-0.629	-0.712	-0.066	103.40	112.10	110.28
R219A+RAl	1.115	0.219	-0.087	-0.636	-0.706	-0.069	104.23	113.15	111.57
R219E+RAl	1.115	0.216	-0.080	-0.636	-0.709	-0.066	104.19	112.35	112.05
R219K+RAl	1.115	0.219	-0.081	-0.635	-0.708	-0.067	104.05	112.70	111.52
WT+RAl	1.115	0.219	-0.082	-0.635	-0.707	-0.069	104.05	112.67	111.56
mPLP,iY									
R219A+SAl	1.115	0.200	-0.030	-0.670	-0.730	-0.041	103.09	114.18	110.59
R219E+SAl	1.114	0.170	-0.043	-0.670	-0.722	-0.047	106.13	111.91	111.28
R219K+SAl	1.114	0.203	-0.038	-0.705	-0.717	-0.031	103.68	112.05	113.38
WT+SAl	1.115	0.191	-0.032	-0.665	-0.731	-0.046	103.14	114.78	110.46
R219A+RAl	1.115	0.205	-0.040	-0.676	-0.726	-0.049	103.73	112.12	112.49
R219E+RAl	1.115	0.211	-0.051	-0.611	-0.744	-0.006	103.85	111.10	111.28
R219K+RAl	1.114	0.201	-0.044	-0.680	-0.731	-0.051	103.70	113.59	110.99
WT+RAl	1.114	0.204	-0.046	-0.676	-0.728	-0.049	103.64	113.65	111.05
iPLP,mY									
R219A+Sal	1.115	0.193	-0.082	-0.602	-0.715	-0.055	103.74	111.95	109.70
R219E+SAl	1.115	0.180	-0.086	-0.673	-0.648	-0.048	104.61	110.69	110.59
R219K+SAl	1.115	0.190	-0.082	-0.599	-0.720	-0.056	103.51	111.88	110.34
WT+SAl	1.115	0.192	-0.086	-0.597	-0.717	-0.059	103.55	111.92	110.10
R219A+RAl	1.115	0.211	-0.106	-0.614	-0.697	-0.063	104.32	113.53	111.32
R219E+RAl	1.115	0.213	-0.108	-0.618	-0.694	-0.063	104.07	111.28	111.12
R219K+RAl	1.115	0.214	-0.107	-0.614	-0.696	-0.064	104.27	113.80	111.18
WT+RAl	1.115	0.211	-0.099	-0.607	-0.705	-0.063	103.95	112.91	111.81
iPLP,iY									
R219A+SAl	1.115	0.197	-0.063	-0.643	-0.736	-0.031	103.84	114.53	110.74
R219E+SAl	1.114	0.160	-0.057	-0.595	-0.760	-0.029	109.96	109.65	109.78
R219K+SAl	1.114	0.188	-0.055	-0.631	-0.737	-0.038	104.06	114.20	110.75
WT+SAl	1.114	0.182	-0.052	-0.634	-0.732	-0.039	104.17	114.47	110.08
R219A+RAl	1.113	0.199	-0.060	-0.644	-0.735	-0.044	104.99	111.39	113.14
R219E+RAl	1.113	0.200	-0.059	-0.642	-0.737	-0.036	104.94	111.41	113.29
R219K+RAl	1.113	0.199	-0.031	-0.660	-0.738	-0.005	105.52	110.29	111.14
WT+RAl	1.114	0.199	-0.030	-0.659	-0.737	-0.006	105.43	110.24	111.12
PLP-K39 complex									
R219A+SAl	1.115	0.181	-0.168	-0.633	-0.677	-0.019	92.06	116.20	108.69
R219E+SAl	1.115	0.173	-0.160	-0.632	-0.681	-0.016	96.34	114.24	109.93
R219K+SAl	1.115	0.180	-0.180	-0.644	-0.657	-0.010	100.99	112.78	106.04
WT+SAl	1.115	0.171	-0.175	-0.630	-0.657	-0.003	101.93	113.34	107.06

O*: carbonyl oxygen; BA1: COO^--αC-NH_3^+; BA2: COO^--αC-Me; BA3: Me-αC-NH_3^+; R-& S-Alanine : free form ionized alanine; mPLP: molecular form PLP, mY: molecular form tyrosine; iPLP: ionized PLP.

When iPLP and mY were used, a strong correlation coefficient was obtained only for the R-alanine carboxyl oxygen (r's = 0.906 and 0.914, respectively). The S-alanine amino hydrogens, but not the nitrogen, were highly correlated with the enzyme activity (r's = 0.973 and 0.868). The locations of the amino and carboxyl groups of the S- and R-alanines were different from that of the PLP aldehyde. The closely located atoms affected the electron transfer. When both PLP and the tyrosines were ionized, the correlation coefficient for S-alanine was not very strong, but the carboxyl oxygen of R-alanine was well correlated (r = 0.952). In the PLP-K39 complex racemase, the S-alanine nitrogen and αC were both well correlated with the reactivity (r's = 0.905 and 0.882 [S→R] and 0.896 and 0.895 [R→S]).

The conformation of the monomer-PLP-alanine complex depended on the initial structure extracted. When only a part of a dimer was used, the contribution of the Y354 hydroxyl and E172 carboxyl groups was important. These groups located near the alanine amino group. A monomer (mPLP, mY) was initially corrected from a downloaded structure, modified to an S-alanine complex, and the complex structure was optimized using MM2. The mutants were then constructed. The alanine carboxyl and amino groups were tightly fixed with the PLP-monomer complex. The contributions of Y354 and E172 were neglected, and the Y43 hydroxyl and D171 carboxyl groups were considered the major electron transfer centers with K39, R139, and PLP aldehyde. The complex conformations were affected by the ionization of PLP and tyrosine from the original molecular form of PLP and tyrosine. Only the Y43 and Y354 hydroxyl groups were ionized. The correlation coefficients for the COO^- - αC - NH_3^+ bond angle generally decreased for complexes with mPLP and mY, mPLP and iY, iPLP and mY, and iPLP and iY (r's = 0.671, 0.680, 0.836, and 0.694 [n's = 4], respectively). The values using $(k_{cat}/K_M)_{R\rightarrow S}$ increased to 0.684, 0.686, 0.841, and 0.705, respectively. The correlation coefficients for R-alanine were not affected. The correlation coefficients for mPLP and mY, iPLP and mY, and iPLP and iY complexes were 0.974, 0.928, and 0.928, respectively, with $(k_{cat}/K_M)_{S\rightarrow R}$; and 0.979, 0.929, and 0.936 with $(k_{cat}/K_M)_{R\rightarrow S}$, respectively.

These correlation coefficients indicated that ionization of the tyrosines did not improve the reactivity. Ionization of both PLP and the tyrosines correlated well with the reactivity of R-alanine, but not S-alanine. Other bond angles that correlated well with the activity of R-alanine were COO^--αC-CH_3 and CH_3-αC-NH_3^+ (r's = 0.959 and 0.961, respectively). When the reactivity in the change from R- to S-alanine was used, the correlation coefficients were 0.962 and 0.965, respectively. The slope was negative, indicating that NH_3^+ was pulled toward the PLP aldehyde and the D171 carboxyl group, and COO^- was pulled toward the K39 ε-amino and R136 guanidino groups. Such a change in the bond angle depended on the possibility of electron transfer through the ionized groups. The ionized alanine carboxyl group was strongly pulled by Coulombic force, especially by the R136 guanidino group and the ionized K39 amino group. The ionized alanine amino group was strongly fixed by the D171 and E172 carboxyl groups. The bond angle of S-alanine did not change significantly, suggesting that polar substitutes of alanine were pulled by Coulombic interaction and/or hydrogen bonding (Lewis acid-base interaction), and the alanine structure became flat, as in the transition state, but the amino group was not replaced.

The values of enzyme activity in the change from S-alanine to R-alanine correlated with the apc of ionic amino hydrogen optimized with mPLP and mY ($r = 0.848$ [$n = 4$]). When the tyrosines were ionized, no practical correlation coefficient was obtained. When iPLP and mY complexes were used, the correlation coefficients between $(k_{cat}/K_M)_{S\rightarrow R}$ with αC, amino nitrogen, and the hydrogen, Y354 oxygen, E171 carbonyl oxygen, or K39 e-amino hydrogen were 0.828, 0.900, 0.950, 0.891, 0.840, and 0.870 ($n = 4$), respectively. When iPLP and iY complexes were used, the correlation coefficients between $(k_{cat}/K_M)_{S\rightarrow R}$ with αC, amino nitrogen and the hydrogen, or Y43 oxygen were 0.866, 0.988, 0.942, and 0.844 ($n = 4$), respectively. The ionization of PLP maintained tight conformation. Several apcs had a large correlation coefficient, but the slope was too small to predict the selective reactivity of the mutants.

The p*K*a of the tyrosine hydroxyl group is 10.07, and therefore tyrosine may maintain its molecular form under physiologic conditions. The D171 and E172 carboxyl groups are ionized, however, and contribute to the reaction. Theoretically, the lower the apc of αH, the higher the reactivity, but the αH apc changed little. The formation of iminium ion adducts between the amino acids and the cofactor PLP results in a large increase in the acidity of the αH [8]. The slopes of the correlation decreased for all conformations, except for R-alanine with both the iPLP and iY. The selectivity of a racemase mutant, therefore, cannot be explained by an increase in the acidity of the α-proton.

The molecular shape of alanine and the apc of target atoms can be used to quantitatively analyze the enzyme reactivity of the racemase mutants. The possible racemization mechanism, however, was not clearly demonstrated. The probable molecular movement was studied using the molecular dynamics program of CAChe. The selected complex was R219K with an R- or S-alanine complex. The PLP was ionized. Molecular dynamics calculations demonstrated the movement of molecules under limited conditions. The complex conformations extracted for the apc calculation were applied to study the conformation changes within a limited time window. Before starting the calculation, the background atoms, the peptide connection atoms, were locked to form a cage to trap the alanine. The calculation time was 2 pico-seconds, and the conformation was evaluated every femtosecond at 310K. The atomic distance from αH to αC or other key atoms should indicate the αH transfer. The bond angle should indicate the maximum flatness. At some point, the alanine becomes carbanion-like. The extracted values are summarized in Table 3.

When a snap shot was taken, the atomic distance between αH and αC had increased only a little, and did not indicate the separation. The total angle expanded more than 20º. At some point, alanine becomes a carbanion without the direct pull of αH from αC. Rather, the electron transfer may occur through the alanine carboxyl group oxygen to form the carbanion. The apc of the oxygen changed considerably compared to that of the other atoms. This mechanism is supported by the high correlation between the mutant reactivity and the oxygen's apc, and between the reactivity and bond angles related to the alanine carboxyl group. It seems, therefore, that the absence of Y265' is not critical.

At the time of the snapshot, the R-alanine αH made contact with the ionized phenolic oxygen of PLP with a minimum atomic distance of 2.365Å, but not with nearby tyrosine residues. Total angles within R-alanine expanded from 318.10 to 344.84°, excluding αH. The atomic distance between αH and αC increased from 1.116Å to 1.139Å.

Table 3. Properties obtained by molecular dynamics calculation

Properties	1st site	2nd site	S-alanine*[1]	R-alanine*[1]
αH-αC distance*[2]	1.133	1.148	1.115	1.116
αH-Y265' hydroxyl oxygen*[3]	2.312	3.117	-	-
αH-PLP aldehyde oxygen*[3]	2.395	4.011	-	-
αH-Y354 hydroxyl oxygen*[3]	-	2.905	-	-
Maximum bond angle*[4]	343.3	347.75	322.34	318.10

*[1]: original values, *[2]: maximum distance Å, *[3]: minimum distance Å, *[4]: maximum bond angle, degree.

The carboxyl oxygen contacted the K39 amino group, and the PLP aldehyde was free from amino groups. According to molecular mechanics calculations, the atomic distances of the optimized conformation between αH and the PLP hydroxyl oxygen, and between the alanine amino group hydrogen and the PLP aldehyde oxygen were 6.141Å and 2.014Å. These distances changed to 2.365Å and 4.968Å, respectively. If this is a real phenomenon inside the racemase, then the contact between the ionized alanine and the ionized PLP is directly affected by R219, and alanine αH moves to the PLP oxygen to form the carbanion structure. The dissociation constant of the hydroxyl group of Y354 is 10.07. Even if the hydroxyl group was not ionized, racemization of R-alanine may occur when αH is close to the ionized oxygen of Y354 with a distance of 2.745Å. The R-alanine carboxyl group is located close to the R136 guanidino group, and the alanine amino group contacted the E172 carboxyl group with atomic distances of 2.979Å and 2.009Å, respectively. The contribution of the PLP aldehyde and hydroxyl groups in this conformation is not clear. The angle of S-alanine expanded from 322.34 to 346.54°, and the atomic distance between αH and αC increased from 1.115Å to 1.154 Å, but αH did not contact either the Y354 oxygen or the PLP oxygen. The racemization mechanism of S-alanine at this location is not clear, and may also occur at a different location. Y265' is located far from PLP, at the edge of the racemaze molecule. The results support those obtained from the dimer extracts.

CONCLUSION

Deprotonation of the αC and the formation of imine are the basic mechanisms underlying alanine racemase reactivity. The atomic distance between αH and αC, and the apc of αH changed little in the conformations of the molecular forms of PLP and tyrosine. The apc of the alanine αC was well correlated with the reactivity, and the value was high for R-alanine. Such a high correlation coefficient was not obtained with ionized tyrosine. When the ionized PLP or the molecular form of the tyrosine complex or both the ionized PLP and tyrosine complex was used, an acceptable correlation coefficient was obtained from the apc of key atoms in both R- and S-alanines. The differences in the apc among mutants and the slopes were small. Therefore, this relation cannot be used for quantitative analysis of enzyme reactions. The alanine carboxyl and amino groups contact the oppositely charged residue's cation and anion for the electron transfer. The bond angles of the substitute alanine were well correlated with mutant reactivity. The bond angle, therefore, can be used for quantitative analysis of the reactivity.

The substrate alanine could not bind with PLP and could not replace K39 bound with PLP; rather, the substrate alanine formed a complex with PLP (for example, R-amino acid oxidase, where the coenzyme FAD was trapped in the R-amino acid oxidase but did not bind with the enzyme). The calculations were performed without solvent (water). The results may not directly compare with those in vitro. The Lewis acid-base interaction is the strongest, however, and because water molecules cannot interrupt the Lewis acid-base interaction, this may work as a bridge to fill the gap. Here, I achieved a quantitative analysis of the selective reactivity of alanine racemase. The R-form of alanine was more stable in complexes with this enzyme due to the favorability of the enzyme structure. The probability depends on the reality of stereo structure. The downloaded secondary structure of pairs of monomers was not the same. The computer system cannot construct the real stereo structure by itself. Which structure is the right structure remains as the basic question. Further study will be carried out how the secondary structure modification affects the enzyme reactivity.

ACKNOWLEDGMENT

I thank Prof. Richard Nigel of Florida University for his critical review of the manuscript.

REFERENCES

[1] Hanai, T. Quantitative in silico chromatography, computational modeling of molecular interactions, *Royal Society of Chemistry,* Cambridge, 2014, pp. 338.

[2] Hanai, T. Quantitative in silico analysis of molecular recognition and reactivity of D-amino acid oxidase, *Internet Electr. J. Mol. Design*, 5 (2006) 247-259.

[3] Hanai, T. Quantitative in silico analysis of enzyme reactions: Comparison of D-amino acid oxidase and monoamine oxidase, *Am. Biotech. Lab*. 25 (2007) 8-13.

[4] Umhau, S.; Pollegioni, I.; Molla, G.; Diederichs, K.; Welte, W.; Pilone, M.S.; Ghisla, S. The x-ray structure of D-amino acid oxidase at very high resolution identifies the chemical mechanism of flavin-dependent substrate dehydrogenation, *Proc. Natl. Acad. Sci. USA*, 2000, 97, 12463-12468.

[5] Smith, G.G.; Sivakus, T. Mechanism of the racemization of amino acids. Kinetics of racemization of arylglycines, *J. Org. Chem.* 1983, 48, 627-634.

[6] Faraci, E.S.; Walsh, C.T. Racemization of alanine by the alanine racemases from Salmonella typhimurium and Bacillus stearothermophilus: *Energetic reaction profiles, Biochemistry*, 1988, 27, 3267-3276.

[7] Watanabe, A.; Yoshimura, T.; Mikami, B.; Hayashi, H.; Kagamiyama, H.; Esaki, N. Reaction mechanism of alanine racemase from Bacillus stearothermophilus, *J. Biol. Chem.*, 2002, 277, 19166-19172.

[8] Toney, M.D. Reaction specificity in pyridoxal phosphate enzyme, *Ach. Biochem. Biophys.* 2005, 433, 279-287.

[9] Major, D.T.; Nam, K.; Gao, J. Transition state stabilization and α-amino carbon acidity in alanine racemase, *J. Am. Chem. Soc.* 2006, 128, 8114-8115.

[10] Jeffrey, P.; Shaw, J.P.; Petsko, G.A.; Ringe, D. Determination of the structure of alanine racemase from Bacillus stearothermophilus at 1.9 Å resolution, *Biochemistry,* 1997, 36, 1329-1342.

[11] Troiani, A.; Gasparrini, F.; Grandinetti, F.; Speranza, M. Chiral ions in the gas phase. 1. Intramolecular racemization and isomerization of O-protonated (S)-trans-4-hexen-3-ol, *J. Am. Chem. Soc.* 1997, 119, 4525-4524.

[12] Sun, S.; Toney, M.D. Evidence for a two-base mechanism involving tyrosine265 from arginine219 mutants of alanine racemase, *Biochemistry,*1999, 38, 4058-4065.

[13] RCSB Protein Data Bank, www.rcsb.org/pdb/.

[14] Sharif, S.; Powell, D.R.; Schagen, D.; Steiner, T.; Toney, M.D. Fogle, E.; Limbach, H-H. X-ray crystallographic structures of enamine and amine Schiff bases of pyridoxal and its 1:1 hydrogen-bonded complexes with benzoic acid derivatives: evidence for coupled inter- and intra-molecular proton transfer, *Acta Cryst.* 2006, B62, 480-487.

In: Molecular Mechanics and Modeling
Editor: Priscilla Watkins
ISBN: 978-1-63483-388-2

Chapter 4

GRAPHENE LAYER PORE MODELS FOR MOLECULE-SURFACE BINDING ENERGIES

Thomas R. Rybolt*[1,*]*, Miranda C. Trentle*[1]*,
Matthew J. Rice*[2]* and Howard E. Thomas*[2]

[1]Department of Chemistry, University of Tennessee
at Chattanooga, Chattanooga, TN, US
[2]Department of Chemistry, Erskine College,
Due West, SC, US

ABSTRACT

Gas-solid chromatography experiments were used to determine the second gas-solid virial coefficient (B_{2s}) values for 10 adsorbate molecules interacting with the solid adsorbent Carbosieve S-III (Supelco). The adsorbates used included: methane, ethane, chloromethane, chlorodifluoromethane, 1,1-difluoroethane, propane, hexafluoroethane, dichlorodifluoromethane, 1-chloro-1,1-difluoroethane, and butane. Carbosieve S-III is a porous carbon powder with a fairly uniform slit width pore size of 0.55 nm and a Nitrogen BET surface area of 995 m^2 /g. The B_{2s} data were collected over a range of temperature and used to construct van't Hoff plots of $\ln(B_{2s})$ versus 1/T for each molecule. The slopes of these plots gave experimental molecule-surface binding energies or adsorption energies (E*). Computational experiments using molecular mechanics with MM2 parameters were used to generate

[*] Corresponding author: E-mail address tom-rybolt@utc.edu (corresponding author T.R. Rybolt).

surface models and calculate the binding energy (Ecal*) and a simple modification of the same (Ecal*$_{modified}$) based on the fraction of nonhydrogen sp^3 carbon atoms in the adsorbate molecule. Ten different models were constructed: plate, parallel plate (D = 0.94 nm, 0.89 nm, 0.84 nm, 0.79 nm, 0.74 nm), and double parallel plate (D = 1.04 nm, 0.99 nm, 0.89 nm, 0.84 nm) where D is the internuclei separation measured from the innermost carbon atoms in the graphene layer of one plate to the innermost carbon atoms of the graphene layer of the other plate. Each plate in the model was constructed of three parallel graphene layers and each graphene layer consisted of 127 hexagonal benzene ring structures. Thus the plate model has one wall with one set of three layers, the parallel plate model has two walls with two sets of three layers, and the double parallel plate model has four walls with four sets of three layers. This last model forms a box-like structure that is open at the top and bottom. Of the 10 models, the double parallel plate model with internuclei parallel wall separation of D = 0.89 nm (interior of 0.55 nm across) correlated best with the experimental molecule-surface binding energies. Experimental values versus these calculated and modified values of molecule-surface binding energies gave a slope of 1.025 with a correlation of R^2 = 0.978. Molecular mechanics based models can provide a representation of average carbon porosity and enable computational estimates of molecule-surface interaction energies.

Keywords: Gas-solid interaction energy, molecular mechanics, molecule-surface binding energy, adsorption energy, graphene, carbon surfaces, nanoporous carbon, Carbosieve S-III

1. INTRODUCTION

Physical adsorption involves interactions between molecules and solid surfaces. Physical adsorption of neutral organic molecules to a carbon surface forms no chemical bonds and has a weak attraction dominated by van der Waals (vdW) dispersive interactions due to transient induced dipoles. Molecule-surface binding energy (E*) also known as gas-solid the interaction energy or adsorption energy is the energy decrease for a gas molecule adsorbed on a solid surface [1]. The solid surface used in this research is Carbosieve S-III (Supelco). This porous carbon powder is fairly uniform with a reported a narrow slit width pore size around 0.55 nm and N_2 BET surface area of 995 m^2/g [1].

Porous materials have many applications in the separation and trapping of molecules. Porous materials are useful for manufacturing, engineering, medicine, environmental monitoring. Some applications of porous carbon powders include gas masks, air filters, water purification, treatment for poisoning, relief from gastric problems, and sewage odor treatment [2]. Porous materials traditionally are divided into three categories: micropores, mesopores, and macropores. Micropores contain pores with a size of less than 2 nm. Mesopores have a range of 2 – 50 nm and macropoes have a pore size greater than 50 nm [3]. Due to the newer focus on nanoscience and nanotechnology, a nanoporous material could be broadly and normally defined as having a dimensionality range of 1 – 100 nm for the pore size. Nanoporous materials are one focus of nanoscience and nanotechnology because these solid surfaces have the ability to adsorb and interact with atoms or molecules on the surface and in the pores [2].

Nanoporosity can be associated with bulk material or membranes. Carbosieve S-III is a bulk material that represents the lower end of nanoporosity due to its small pore size. Carbosieve S-III powder has found a variety of uses including: as a means of air sample collection [4], as a model for microporous adsorption of hydrogen and carbon dioxide [5], for nitropropane determination in workplace air [6], for a gas-chromatographic method for determination of volatile organic compounds [7], and for analysis of odorous organic compounds [8].

B_{2s} represents the second gas-solid virial coefficient and is an experimental value obtained from the low-coverage Henry's law region of adsorption. As such it reflects isolated molecule-surface interactions where adsorbate-adsorbate interactions can be ignored. The amount of gaseous molecules adsorbed on a surface or in a pore or retained in a chromatographic experiment depends on molecule-surface binding energies (interaction or adsorption energies) and surface structures. A higher molecule-surface binding energy increases the value of B_{2s} and this in turn will increase the amount of gas adsorbed on a surface or alternatively increase the length of time for a dose of gas to pass through a powder chromatographic column [9].

In this work the experimental molecule-surface binding energy (E*) is obtained from the temperature dependence of B_{2s} values using van't Hoff plots of $\ln(B_{2s})$ versus 1/T, where T is the temperature. The experimental work utilized 111 different B_{2s} values (new presented here and prior) over a range of temperatures for 10 different adsorbate organic molecules [1]. The ten gas molecules used for this research were (in order of increasing E*) methane; ethane; chloromethane; chlorodifluoromethane; 1,1-difluoroethane; propane;

hexafluoroethane; dichlorodifluoromethane; 1-chloro-1,1-difluoroethane; and butane. The B_{2s} and T data pairs for these organic molecules were used to find the 10 experimental molecule-surface binding energies.

Ten different model surfaces with various representations of porosity were prepared and used with MM2 calculations to determine molecule-surface binding energies. A comparison of the experimental and computed values allowed selection of an appropriate surface representation and presented insight into nature of the surface. In addition to extracting some basic surface structural information, this approach allows simple molecular mechanics calculations to be used to predict binding energies for molecules where experimental values are not available.

Molecular Mechanics, B_{2s} equations, and gas-solid chromatography experimentation all relate to one another in this research of adsorbate-adsorbent interactions. Molecular mechanics was used to construct different models of plate, parallel plate (D = 0.94 nm, 0.89 nm, 0.84 nm, 0.79 nm, 0.74 nm), and double parallel plate (D = 1.04 nm, 0.99 nm, 0.89 nm, 0.84 nm) where D is the internuclei separation measured from the innermost carbon atoms in the graphene layer of one plate to the innermost carbon atoms of the graphene layer of the opposite side plate. These models are utilized to calculate the binding energy (Ecal*) and a modified calculated binding energy $Ecal^*_{modified}$. Comparisons of E* and Ecal* and E* and $Ecal^*_{modified}$ can be made in order to correlate experimental and force field computational work. From these comparisons, correlations and predictions of molecule-surface binding energies are possible.

Molecular mechanics force fields [10-12] have been widely used for molecular conformation studies [13, 14]. They have also been used to estimate molecule-carbonaceous surface binding energies [1, 15-17]. Dispersion or van der Waals (vdW) forces dominate the adsorption of neutral organic molecules on a carbon surface, and in prior work we have shown that existing MM2 parameters are well suited to find calculated molecular binding energies for porous, rough, and smooth carbon surfaces [1, 15-17]. Calculated, Ecal*, and experimental, E*, molecule-surface binding energies were found to be in close agreement in these earlier studies and they have been summarized previously [18]. Other prior work using MM2 parameters estimated the nucleobase binding energies for a single graphene layer using to be 0.704, 0.639, 0.579, 0.579, and 0.509 eV for guanine, adenine, thymine, cytosine, and uracil, respectively [18].

A variety of prior research described the importance of carbon, adsorption and the pore structure of materials [19 - 25]. One approach is Molecular

Mechanics (MM) which is an approximated calculation of the energy of a molecule using force fields. The energy is made up of components such as: bond energy, bond angle energy, vdW or dispersion interactions, torsion angle energy, charge-charge interactions, and miscellaneous. The energy that is most prominent in MM2 calculations for the energy change in a neutral molecule attracted to a carbon surface is the van der Waals interaction [26].

Different thermodynamic techniques, such as gas-solid chromatography and thermal programed desoprtion, have been used to identify the physical properties of adsorbents [18, 27]. Along with this experimental research, computational research ranges from theoretical grand canonical Monte Carlo simulations of graphene oxide framework (GOF) materials for H_2 storage [19] to adding energetic heterogeneity to the pore walls of a standard slit pore model [23]. Different types of molecular modeling have been used. For example, two classes of models (quasi-1D carbon nanotubes and 2D single-layer carbon membranes) were computed in order to obtain three types of separation mechanisms: size sieving, quantum sieving, and chemical affinity sieving [22].

In another study a model of graphene layers where a parallel pair of finite-length walls were perpendicular to the z-axis was used [24]. A simulation of a flat graphene slit pore was computed between hydrogen molecules and graphitic walls. The model consisted of a carbon slit pore that contains two parallel flat graphene layers at a distance calculated by density functional theory (DFT) and second-order Moller-Plesset (MP2) [21]. The slit pore model may be constructed from parallel graphene layers. The influence of the number of graphene layers in pores walls on alkane adsorption has also been studied [28].

Second gas-solid virial coefficients and how these measures of molecule-surface interaction are related to molecule-surface potential energies are discussed in the theory section. How experimental values of molecule-surface binding energy are determined from gas-solid chromatography and van't Hoff plots are discussed in the experimental section. Theory and experiment have been related in the past for some carbon surfaces by assuming a Lennard-Jones potential form for molecule-surface interaction and varying the potential well depth until the calculated B_{2s} values with appropriate integral expressions best matched the experimental ones over a range of temperatures for each adsorbate [9]. However, another approach to generate calculated values is used in this work. Specifically, standard MM2 parameters and molecular mechanics are used to find molecule-surface binding energies.

Figure 1. Plate model (P) with ethane gas on the surface. Flat plate model consists of 127 hexagonal rings per layer with three layers of graphene.

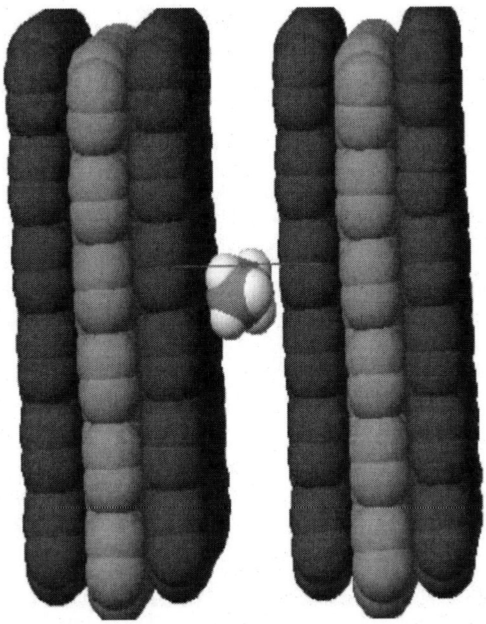

Figure 2. Parallel plate model (PP) with ethane gas in the pore. Parallel plate model consists of two plates set at a distance (D) representing the internuclei separation measured from the innermost carbon atoms in the graphene layer of one plate to the innermost carbon atoms in the graphene layer of the other plate.

Three basic surface models are created and compared in this approach. The models vary from a flat surface to a pore made of two parallel surfaces to a box-like pore with four surfaces. The three basic surface adsorbent models with inclusion of adsorbate molecule ethane are shown in Figures 1, 2, and 3. The Figure 3 pattern provided a convenient way to enlarge or reduce the rectangular box without modifying the size of the side plates. The objective of this work is to develop an appropriate surface model that with standard force field calculations can be used to generate molecule-surface binding energies closely matching experimental values. By matching experimental values, it is expected that useful predictions could be made for other molecules on the same carbon surface.

2. THEORY

The second gas-solid virial coefficient, B_{2s}, is defined as the measure of interaction of a gas molecule on a solid surface. In the limit of low coverage, the relationship between the amount of gas adsorbed, n_{ads} (mol/g), and B_{2s} (cm^3/g) is given as

$$n_{ads} = \frac{B_{2s}}{RT} P \tag{1}$$

where R (82.06 cm^3atm/mol K) is the gas constant, T (K) is the temperature, and P (atm) is the pressure [9].

B_{2s} may be represented by the experimental chromatographic equation

$$B_{2s} = t\,F_1/m \tag{2}$$

where t is the corrected retention time (s), F_1 is the corrected flow rate (cm^3/s) through the column, and m is the mass of the solid surface (g). The retention time (t) is defined as the amount of time beyond the no interaction flow rate time it takes for the gas to make its way through a column packed with the carbon solid of interest. This time is dependent upon the interaction between the gas molecules and the packed solid carbon. For Eq. (2) there must be corrections made to the measured retention time and measured flow rate. The gas retention time is expressed as

$$t = t_s - t_m \tag{3}$$

where t_s is the time from injection to detection of the sample gas and t_m is the time of injection to detection of the marker gas that has negligible gas-solid interaction [29] at the temperature of interest.

The corrected flow rate is calculated by the equation

$$F_1 = 1.5\, F_m \left(T_1/T_f\right) (P_o/P_a)\, [(P_i/P_o)^2 - 1]/[(P_i/P_o)^3 - 1] \tag{4}$$

where F_m is measured flow rate, T_1 is the column temperature, T_f is the temperature of the flow meter, P_o is the outlet pressure, P_a is the atmospheric pressure, and P_i is the inlet pressure [9].

B_{2s} may also be represented by the theoretical equation below and calculated by using an integral expression that is dependent upon the gas-solid interaction potential, u_{1s}. The formula for B_{2s} given as a function of u_{1s} is

$$B_{2s} = \int \left(e^{(-u_{1s}/kT)} - 1\right) dV \tag{5}$$

where k is the Boltzmann constant, T is the temperature in kelvin, and dV is the volume element [30]. A B_{2s} equation suitable for a flat plate model, parallel plate model or other more complicated porous models can be expressed in general as

$$B_{2s} = \iiint \left(e^{(-u_{1s}/kT)} - 1\right) dr_1\, dr_2 dz. \tag{6}$$

The three axes (r_1, r_2, and z) correspond to the Cartesian coordinate system where r_1 and z are parallel to the surface. The axis z can be considered vertical (going down into the pore) and r_1 a left and right orientation if the observer is facing the surface forming one plate of the pore. The r_2 axis is perpendicular to the surface.

2.1. Flat Plate Model

To adapt the equation to a flat surface model, the equation is expressed as

$$B_{2s} = \int dr_1 \int dz \int \left(e^{(-u_{1s}(r_2)/kT)} - 1\right) dr_2. \tag{7}$$

The expression can be reduced to

$$B_{2s} = A \int_0^{\infty} \left(e^{(-u_{1s}(r_2)/kT)} - 1\right) dr_2 \tag{8}$$

where A is expressed as the total surface area of the solid with units of cm^2/g. The final expression for B_{2s} given in cm^3/g and based on a single uniform, flat surface called a plate model (P) might be expressed as

$$B_{2s} = A\, r_2^* \int_0^{10} \left[e^{(-u_{1s}(y)/kT)} - 1\right] dy \tag{9}$$

where r_2* is the equilibrium distance of the gas molecule from the solid surface, y is a reduced variable ($y = r_2/r_2$*), $dy = dr_2/r_2$*, and r_2 is the distance from the center of an adsorbate molecule to the solid surface atoms along an axis normal to the surface. Integral limits of y = 0 to 10 are sufficient since if y = 10 then r = 10 r_2* so the interaction potential is ten times the equilibrium separation.

The interaction potential, u_{1s}, for a flat surface can be given as

$$\frac{u_{1s}(y)}{k} = E^* \left\{\left[\frac{n}{(n-m)}\right]\left[\left[\frac{n}{m}\right]^{\left[\frac{m}{(n-m)}\right]}\right]\right\}\left\{y^{-n}\left[\frac{n}{m}\right]^{\left[\frac{-n}{(n-m)}\right]} - y^{-m}\left[\frac{n}{m}\right]^{\left[\frac{-m}{(n-m)}\right]}\right\} \tag{10}$$

where E^* is the interaction energy in kelvin, n is the repulsive parameter, and m is the attractive parameter. Eq. (10) is substituted into Eq. (9) to evaluate the integral. Evaluation of the integral would be over the limits from $r_1 = 0$ to ∞ but from a practical standpoint $r_2 = 0$ to 10 r* or y = 0 to 10 are sufficient.

2.2. Parallel Plate Model

From Eq. (6), B_{2s} for a parallel plate model (PP) can be expressed as

$$B_{2s} = \int dr_1 \int dz \int_{-a}^{a} \left(e^{(-u_{1s}(r_2)/kT)} - 1\right) dr_2 \tag{11}$$

where r_1, r_2, and z are variables in the Cartesian coordinates, and a is half the distance between the two plates. Therefore +a is the upper limit and –a is the lower limit of the integral in Eq. (11) with 0 being the point half way between the two surfaces.

Using the equation

$$A = (N_{pores})(A_{pore}) \tag{12}$$

where A is the total area of all the pores, N_{pores} is the number of pores, and A_{pore} is the area of a given pore formed from two parallel plates measured in m^2/g or cm^2/g. B_{2s} can be expressed as

$$B_{2s} = N_{pores} \int_0^{L_{r_1}} dr_1 \int_0^{L_z} dz \int_{-a}^{a} \left(e^{(-u_{1s}(r_2)/kT)} - 1\right) dr_2 \tag{13}$$

where L_{r_1} is the upper limit in the r_1-direction and L_z is the upper limit in the z-direction. The equation for the area of the pore can be substituted into the equation where

$$B_{2s} = N_{pores} A_{pore} \int_{-a}^{a} \left(e^{(-u_{1s}(r_2)/kT)} - 1\right) dr_2. \tag{14}$$

The equation can then be expressed as

$$B_{2s} = A \int_{-a}^{a} \left(e^{(-u_{1s}(r_2)/kT)} - 1\right) dr_2. \tag{15}$$

By dividing and multiplying the function by a, an expression with a reduced variable can be used so that B_{2s} with units of cm^3/g is given by

$$B_{2s} = A\, a \int_{-a}^{a} \left(e^{(-u_{1s}(r_2)/kT)} - 1\right) \frac{dr_2}{a} \tag{16}$$

or

$$B_{2s} = a\, A \int_{-1}^{+1} [e^{(-u_{1s}(y)/kT)} - 1]\, dy \tag{17}$$

where $y = \frac{r_2}{a}$ and since r_2 varies from –a to +a then the limits of y are -1 to +1.

Due to symmetry considerations

$$B_{2s} = 2a\, A \int_0^{+1} [e^{(-u_{1s}(y)/kT)} - 1]\, dy \tag{18}$$

where y is a reduced variable which is equal to $y = r_2/a$. The equation can be given as

$$B_{2s} = 2a N_{pores} A_{pore} \int_0^1 \left[e^{(-u_{1s}(y)/kT)} - 1\right] dy. \tag{19}$$

This can be written as an expression of volume so that

$$B_{2s} = N_{pores}V_{pore}\int_0^1\left[e^{(-u_{1s}(y)/kT)} - 1\right]dy \tag{20}$$

where $2aA_{pore}$ is equal to V_{pore} which is equal to the volume of a single pore. Therefore,

$$B_{2s} = V_{PP}\int_0^1\left[e^{(-u_{1s}(y)/kT)} - 1\right]dy \tag{21}$$

where V_{PP} is equal to $2aN_{pores}A_{pore}$ and is the total volume of all the parallel plate model pores. So total pore volume is V_{PP} or $N_{pores}V_{pore}$ and V_{pore} is the volume of a single pore.

To simplify notation to be used below, r is used in place of r_2 for remaining PP equations. For the parallel plate model, the molecule is placed between the two parallel plates forming a pore with no sides. One distance is expressed as (a - r) and the other (a + r) where a is half the distance between the two plates and r is the distance from the molecule to the center along a line perpendicular to the two plates. The two interaction potentials for the parallel plate model can be expressed as

$$u_{1s}(a-r)/k = E^*p_1\{p_2[r^*/(a-r)]^n - p_3[r^*/(a-r)]^m\} \tag{22}$$

$$u_{1s}(a+r)/k = E^*p_1\{p_2[r^*/(a+r)]^n - p_3[r^*/(a+r)]^m\} \tag{23}$$

where the reduced parameters p_1, p_2, and p_3 can be expressed as

$$p_1 = [n/(n-m)][n/m]^{[m/(n-m)]} \tag{24}$$

$$p_2 = [n/m]^{[n/(m-n)]} \tag{25}$$

$$p_3 = [n/m]^{[m/(m-n)]}. \tag{26}$$

Once the interaction potentials are added together, the equation can be expressed as

$$\frac{u_{1s}(y)}{k} = E^*p_1\{p_2([\theta^*/1-y]^n + [\theta^*/1+y]^n) - p_3([\theta^*/1-y]^m + [\theta^*/1+y]^m)\} \tag{27}$$

where $\theta^* = \frac{r^*}{a}$ and for example $\frac{r^*}{a+r} = \frac{r^*/a}{a/a+r/a} = \frac{\theta^*}{1+y}$. In order to evaluate the integral, Eq. (27) is plugged into Eq. (17) or (18) and evaluated over the limits expressed in the respective equations. Recall that r* is the molecule-surface equilibrium separation.

2.3. Double Parallel Plate Model

The double parallel plate model (DPP) repeats the parallel plate model by adding another pair of plates to form a square or rectangle. For this model, the interaction energy potential can be expressed as

$$\frac{u_{1s}(y)}{k} = E^* p_1\{p_2([\theta_2^*/1-y]^n + [\theta_2^*/1+y]^n) - p_3([\theta_2^*/1-y]^m + [\theta_2^*/1+y]^m)\} \quad (28)$$

for $\theta_2^* = \frac{r_2^*}{a_2}$ and $y = \frac{r_2}{a_2}$ for the front to back direction (imagine facing the surface with the molecule on it) and

$$\frac{u_{1s}(x)}{k} = E^* p_1\{p_2([\theta_1^*/1-x]^n + [\theta_1^*/1+x]^n) - p_3([\theta_1^*/1-x]^m + [\theta_1^*/1+x]^m)\} \quad (29)$$

where $\theta_1^* = \frac{r_1^*}{a_1}$ and $x = \frac{r_1}{a_1}$ for the left to right direction.

From the above interaction energy potential equations, the B_{2s} equation for the double parallel plate model is

$$B_{2s} = \int_0^{L_z} dz \int_{-a_1}^{+a_1} \left(e^{(-u_{1s}(x)/kT)} - 1\right) dx \int_{-a_2}^{+a_2} \left(e^{(-u_{1s}(y)/kT)} - 1\right) dy. \quad (30)$$

The above equation is for a rectangle with the wall separation distance of $2a_1$ and $2a_2$ and a pore depth of L_z. If the plates for the DPP model form a square ($a_1 = a_2 = a$), then the equation may be simplified to

$$B_{2s} = L_z 2a \int_0^1 \left(e^{(-u_{1s}(x)/kT)} - 1\right) dx \; 2a \int_0^1 \left(e^{(-u_{1s}(y)/kT)} - 1\right) dy. \quad (31)$$

where $y = \frac{r_2}{a}$ and $x = \frac{r_1}{a}$.

Since the two integrals should be equivalent then, B_{2s} can also be expressed as

$$B_{2s} = 4L_z a^2 \int_0^1 \left(e^{(-u_{1s}(y)/kT)} - 1\right)dy \int_0^1 \left(e^{(-u_{1s}(y)/kT)} - 1\right)dy \quad (32)$$

where the volume of a single pore (V_{pore}) is equal to $4L_z a^2$. Therefore, B_{2s} is given as

$$B_{2s} = N_{pores} V_{pore}\, 2 \int_0^1 \left(e^{(-u_{1s}(y)/kT)} - 1\right)dy \quad (33)$$

or

$$B_{2s} = V_{PP}\, 2 \int_0^1 \left(e^{(-u_{1s}(y)/kT)} - 1\right)dy \quad (34)$$

where N_{pores} is the number of pores and (V_{PP}) is the total volume of all the DPP model pores.

The chosen integral representation for P, PP, or DPP can be used to calculate B_{2s} by substituting an Ecal* and varying until the "best" B_{2s} values are found by comparison for a set of molecules with a series of B_{2s} values over a range of temperature. Other parameters within the integral expression must be selected or varied to find the best fit. An iterative approach can be used to find the Ecal* along with other parameters in the integral equation that is chosen. This approach has been described in detail elsewhere [9]. Although the plate, parallel plate, and double parallel plate models could be represented in terms of the equations above, molecular mechanics provides a simpler approach in that surface models can be built to represent the surfaces of interest and direct calculations of Ecal* can be done using existing molecular mechanics parameters. The equations above are presented to show how models could be developed but the focus will be on the molecular mechanics approach.

As indicated in Eq. (1) in the low coverage Henry's Law region, a plot of the moles of gas adsorbed, n_{ad}, versus pressure gives a linear isotherm with a slope equal to B_{2s}/RT where R is the gas constant and T is the kelvin temperature. As the temperature is decreased then the value of B_{2s} is increased, indicating a greater gas-solid interaction relative to the available thermal energy. B_{2s} is given by an integral that is dependent on u_{1s}, the gas-solid interaction potential [9]. This potential in turn depends on several factors including the depth of the gas-solid potential function shown in Eq. (10).

In the current work, the experimental gas-solid interaction energy or binding energy is given by E*. A determination of this energy is the from experimental B_{2s} data using van't Hoff plots of $\ln(B_{2s})$versus 1/T. The experimental binding energy (E*) is equal to the slope in kelvin of the van't Hoff plot.

Table 1. Second gas-solid virial coefficients (B_{2s}) for Carbosieve S-III adsorbent from gas-solid chromatography at various temperatures

	B_{2s} (cm^3/g)	T (K)
methane CH_4	16.5	373
	14.2	383
	12.1	393
	10.5	403
	9.2	413
ethane C_2H_6	81.5	428
	65.5	438
	53.9	448
	44.3	458
	37.7	468
	31.6	478
chloromethane CH_3Cl	102.7	463
	83.2	473
	70.1	483
	57.9	493
	35.9	523
	30.2	533
	25.4	543
	21.9	553
chlorodifluoromethane $CHClF_2$	60	493
	48.5	503
	40	516
	32.5	523
	26.7	533
1,1-difluoroethane $C_2H_4F_2$	60	483
	50	493
	40	503
	32.4	513
	26.1	523

	B_{2s} (cm^3/g)	T (K)
propane C_3H_8	77.1	523
	63.3	533
	53.5	543
	43.8	553
hexafluoroethane C_2F_6	61	363
	42	373
	25	383
	14	393
dichlorodifluoromethane CCl_2F_2	27.7	463
	21.1	473
	14.2	483
	11.3	493
1-chloro-1,1-difluoroethane $C_2H_3ClF_2$	47	503
	39	513
	26	523
	22	533
	16.4	543
	12.4	553
butane C_4H_{10}	121.4	583
	93.1	593
	79.7	603
	63.4	613
	57.7	619

3. Experiment

Gas-solid chromatography was used to find second gas-solid virial coefficients B_{2s} over a range of temperatures (see Table 1). The column packing solid used was Carbosieve S-III powder (Supelco) which has fairly uniform pores. This carbon molecular sieve has a slit width pore size of 0.55 nm and a N_2 BET surface area of 995 m^2/g [29]. The ten adsorbates used included methane, ethane, chloromethane, chlorodifluoromethane, 1, 1-difluoroethane, propane, hexafluoroethane, dichlorodifluoromethane, 1-chloro-1,1-difluoroethane, and butane. In the experiment, the gas molecule along with neon acting as a marker gas were co-injected into a column packed with the solid, Carbosieve S-III. An indication of the Henry's Law interaction between isolated gas molecules and the solid is expressed in terms of B_{2s} measured in cm^3/g and represents an equilibrium balance between adsorbed and free molecules at a given temperature. Van't Hoff plots of $\ln(B_{2s})$ versus 1/T where were T is the kelvin temperature were used to obtain the molecule-surface binding energies.

Determination of B_{2s} values required the mass (g) of the solid adsorbent used in the packed column, the retention times (s) which were corrected for transit time from injection to detector based on a relatively noninteracting neon marker gas, and corrected flow rates (cm^3/s) that included adjustments for column temperatures and pressure drops across the column [9]. Measurements used a Hewlett-Packard 5890 GC with a thermal conductivity detector and an H-P 3392A integrator.

The experimental procedure has been used in previous studies and described in detail [9]. For this work, the 46 x 0.318cm stainless steel column was packed by Supelco Inc. and contained 1.243 g of 60/80 mesh Carbosieve S-III. The column was outgassed in a stream of helium at 623K for at least 12 hours before each day's runs. Helium flow rates were measured with a bubble flow meter and ranged from 0.51 to 0.63 mL/s. The reference helium flow through the detector was switched off before flow rates were measured. Inlet gas pressure ranged from 43 to 97 kPa. Adsorbates were co-injected with the marker gas, neon, using a Hamilton gas syringe. At low enough injection sizes, values of B_{2s} were found to no longer vary with sample size. This consistency indicated Henry's law region of low coverage with molecule-surface interaction dominating over (molecule-molecule)-surface. A series of multiple injections with multiple B_{2s} determinations were averaged to determine B_{2s} at each different temperature. Data for a typical adsorbate covered a range of 30 to 90 degrees within the temperature range 363 to 619K.

Second gas-solid virial coefficient, B_{2s}, values were calculated using Equations 2, 3, and 4 given above. Additional experimental details are described elsewhere [9]. The binding energy, E*, for each molecule was found from plots of $\ln B_{2s}$ versus 1/T. As shown by the R^2 values in Table 2, these plots are quite linear. The reported E* values are taken directly from the slope of $\ln(B_{2s})$ versus 1/T. Kelvin units may be retained for these binding energy values (equivalent to energy because of implicit division by the Boltzmann constant).

Table 2. Experimental values for molecule-surface binding energy from plots ln (B_{2s}) versus 1/T from current and prior (B_{2s},T) data

Name	Formula	Data Points	Slope (K)	Intercept	R^2
methane	CH_4	15	2287	-3.321	0.9997
ethane	C_2H_6	18	3859	-4.625	0.9996
chloromethane	CH_3Cl	20	4356	-4.775	0.9994
chlorodifluoromethane	$CHClF_2$	9	5255	-6.553	0.9895
1,1-difluoroethane	$C_2H_4F_2$	5	5297	-6.852	0.9974
propane	C_3H_8	16	5402	-5.982	0.9984
hexafluoroethane	C_2F_6	4	7021	-15.163	0.9861
dichlorodifluoromethane	CCl_2F_2	8	7038	-11.868	0.9916
1-chloro-1, 1-difluoroethane	$C_2H_3ClF_2$	6	7490	-11.004	0.9909
butane	C_4H_{10}	10	7522	-8.155	0.9861

The B_{2s} data for the same gas were normally collected on multiple days with six to eight repeats at each temperature, and for each temperature the average was taken. This average of six to eight injections is considered as one data point. The B_{2s} data are listed in Table 1 for the 10 adsorbates. The experimental molecule-surface binding energy or adsorption energy (E*) for each molecule was then found by plotting the natural logarithm of B_{2s} values versus 1/T (1/K) [31]. The new data reported here (Table 1) were combined with our prior published data to give a larger set of B_{2s} and T data points [1].

Table 2 provides a list of these van't Hoff plot results based on ln(B_{2s}) versus 1/T for the total data set. For each of the 10 molecules, the number of data points, r^2, slope (K), intercept, and E*(kcal/mol) values are given. As expected we observe that more polarizable molecules have stronger interactions. For example, going from methane to ethane to propane to butane gives E* of 2287, 3859, 5402, and 7522 K, respectively.

Kelvin units can be converted to kcal/mol by dividing E*(K) by 503.22 (K/(kcal/mol)). Larger E* values indicate that an isolated molecule is more strongly attracted to the surface under investigation. E* values incorporate information about complex gas-solid adsorption energy potentials and surface structures [9]. When molecules undergo a transition from the lower potential energy of the adsorbed phase to the higher potential energy of the gas phase there must be an input of energy. Therefore, desorption energies are positive and adsorption energies are negative. However, the E* values are often reported as positive absolute values with the understanding that the sign depends on the process of interest.

4. Analysis and Results

One method of analysis is to utilize B_{2s} and temperature experimental values and find the appropriate force field potential well depths expressed in the flat plate (P), parallel plate (PP) and double parallel plate (DPP) model using B_{2s} integral equations given previously. In that approach the correct well depth for a given model is the one that generates calculated B_{2s} values for a given molecular adsorbate that most closely match the experimental B_{2s} values over a range of temperatures. That approach has been used previously [15-17] with other adsorbents and various models and normally involves an iterative approach to try a series of E* values that give the best calculated B_{2s} match to experimental B_{2s}.

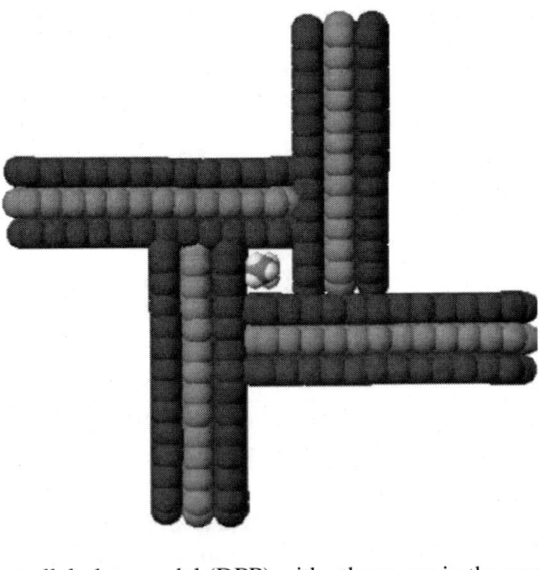

Figure 3. Double parallel plate model (DPP) with ethane gas in the pore. Double parallel plate model consists of two separate sets of parallel plates positioned perpendicular to each other forming a square in the center of the model.

A more direct method, and the one presented in this work, takes advantage of the utility of the force field parameters from MM2 molecular mechanics and their ability to correctly represent van der Wasls (vdW) interactions for neutral organic molecules on model carbon surfaces. This method was used to calculate the binding energy (Ecal*) for the ten adsorbate molecules using the MM2 parameters within Scigress (Fujitsu Version 7.7.0.49) software. Three distinct models were used in which model representations were created for the flat plate (P), parallel plate (PP), and double parallel plate (DPP) surfaces. These three basic surface adsorbent models are shown in Figures 1, 2, and 3. Each figure includes an ethane adsorbate molecule. The Figure 3 pattern provided a convenient way to enlarge or reduce the rectangular box without modifying the any wall plate size.

The flat plate model consisted of three parallel layers of graphene each consisting of 127 benzene carbon rings with an a-b-a packing arrangement so that the carbon atoms of alternate layers are directly underneath each other. In this and the two other models, one plate consists of these three graphene layers of 127 rings each. For the plate model, there is one flat plate that represents a uniform surface. The other two models each consist of multiples of this single plate.

The parallel plate model (PP) consists of two plates, where again each plate is made of the three parallel layers of graphene. These two plates are parallel to each other at a distance (D). D is the internuclei separation measured from the innermost carbon atoms in the graphene layer of one plate to the innermost carbon atoms of the graphene layer of the other plate. The distance can also be described as the pore diameter (d) that is the available empty distance between the plates. The relationship between the internuclei separation (D) and the pore diameter (d) is described by the equation

$$D = d + 0.34 \text{ nm} \qquad (35)$$

where 0.34 nm is the diameter of a carbon atom with the vdW carbon radius of 0.17 nm.

The third model used is the double parallel plate model (DPP) that consists of two separate parallel plate models. These separate sets of parallel plates are set perpendicular to each other. Each plate can be shifted in the up and down direction for the vertical plates (| |) and in the left and right direction for the horizontal plates (==). Looking down onto the plane of the page, is equivalent to looking into the opening of the pore. The plates are aligned to form a rectangular shape. However, in this work, the separation of the plates perpendicular and horizontal were set equal so the rectangular shape was a square shaped box with 4 walls and no top or bottom. The pairs of plates are the same distance apart in all DPP models used.

For both the parallel plate and the double parallel plate (limited to square pore arrangement), different pore diameters (D) were constructed in order to find the model that correlated best with the experimental results. A total of 10 models were used including: P, PP (D = 0.94 nm), PP (D = 0.89 nm), PP (D = 0.84 nm), PP (D = 0.79 nm), PP (D = 0.74 nm), DPP (D = 1.04 nm), DPP (D = 0.99 nm), DPP (D = 0.89 nm), and DPP (D = 0.84 nm). In summary there is one single plate model, five parallel plate pore models with one set of sides and no top or bottom, and four square double parallel plate model pores with no top or bottom. Note that in going from P to PP and to DPP, the wall interaction increases. As the wall separation is decreased, then the molecule-surface porosity interaction is increased. The P model is a nonporous model and the PP and DPP are porous models.

For these calculations, the atoms that have the largest van der Waals radius are given preference to face the solid surface when placing the gas molecule on the graphite surface. The gas molecules are placed in the center of the surface. For example for methane the central atom is placed above the center carbon ring of the first graphene layer with three hydrogen atoms "touching" the surface. For chloromethane the chlorine atom and two hydrogen atoms would be oriented toward the surface. For larger molecules such as ethane, the molecule is placed over the top layer of graphene so that the carbon-carbon bond is centered over the center carbon ring. Each molecule is placed so that the atoms that are closest to the surface are slightly overlapping with the surface so that the molecule is "too close" to the surface. This allows the geometry optimization calculation to push out the molecule to the optimal interactive distance away from the surface and thus reduce the repulsive vdW forces and increase vdW attractive forces.

For the PP and DPP models, one plate is chosen for the molecule to be closest to so that the molecule can be pushed out toward the center of the pore. The molecule position is determined by the sum of the forces for the molecule's atoms and the carbon atoms making up the pore. A default VdW cut off distance of 0.90 nm was used as it has been shown to give computed Ecal*results in good comparison to experimental E* values in prior work [32].

The binding energy measured in kcal/mol is calculated using the equation

$$Ecal^* = | E_{sm} - (E_m + E_s) | \tag{36}$$

where E_{sm} is the energy with the molecule placed on the surface of a plate of the model of interest, E_m is the energy of the isolated molecule, and E_s is the energy of the isolated surface model. E* is reported in kelvin and Ecal* is measured in kcal/mol so the conversion factor 1kcal/mol = 503.2 K is used to convert Ecal* to kelvin. The calculated binding energy is a negative value; however it is expressed as an absolute value or positive value in units of kelvin in order to more easily compare the calculated binding energy with the experimental binding energy. Table 3 gives the results of the Eq. (3) calculation for a double parallel plate model when D = 0.89 nm and lists the 10 adsorbate molecules.

Notice that Table 3 shows the steric energies of each isolated molecule, of the surface model, and of each molecule on the surface. If the difference between the sum of the molecule and surface energies subtracted from the molecule-surface energy was zero (Ecal* = 0), then this would indicate that there was no interaction between the surface and molecule. However, what we observe from the calculations based on Eq. (36) is a stabilization primarily due to favorable vdW interactions between the molecule and the surface. Formally Ecal*(kcal/mol) for adsorption would be negative but are presented as absolute values for Ecal*(K) that are shown in Table 3.

It is important to note that for molecular mechanics calculations there is not a meaning associated with comparing the steric energy of different isolated molecules. However, as the conformation of one molecule changes, and the steric energy decreases then the molecule is exhibiting an energetically more stable configuration. So as a molecule is moved toward a carbon surface there will be a lower collective molecule and surface energy due primarily to vdW forces of attraction. After going through the most favorable interaction energy minimum, if the molecule is moved still closer to the surface carbon atoms, the collective molecule and surface energy will rise rapidly as atoms are forced too close together and vdW forces of repulsion dominate. The values in Table 3 represent the most favorable interaction energy and thus Ecal* is the computed molecule-surface binding energy.

The detailed calculation for the surface DPP (D = 0.89nm) presented in Table 3 was repeated for each of the other nine surface models. For the 10 molecules and 10 surfaces those calculations generated a total of 100 different values of Ecal*. Those 100 values of Ecal* are shown in Table 4. The last column in Table 3 is therefore one of 10 columns of Ecal* values in Table 4. Prior to addressing the question of the best model, there is a simple modification of the Ecal* values that must be done.

Table 3. Molecule-surface binding energy Ecal* using MM2 molecular mechanics for DPP model when D = 0.89 nm where steric energy for molecule (E_m), surface E_s, and molecule with surface E_{ms} are reported

Name	Formula	E_m	E_s	E_{ms}	Ecal* (kcal/mol)	Ecal* (K)
methane	CH_4	0.000	-5082.13	-5087.94	-5.81	2923
ethane	C_2H_6	0.816	-5082.13	-5090.57	-9.25	4657
chloromethane	CH_3Cl	0.118	-5082.13	-5091.35	-9.34	4700
chlorodifluoro methane	$CHClF_2$	0.003	-5082.13	-5093.08	-10.95	5513
1,1-difluoroethane	$C_2H_4F_2$	0.727	-5082.13	-5092.13	-10.73	5401
propane	C_3H_8	1.501	-5082.13	-5093.89	-13.26	6672
hexafluoroethane	C_2F_6	38.610	-5082.13	-5057.58	-14.06	7074
dichlorodifluoro methane	CCl_2F_2	0.052	-5082.13	-5096.76	-14.68	7390
1-chloro-1, 1-difluoroethane	$C_2H_3ClF_2$	0.804	-5082.13	-5095.86	-14.54	7315
butane	C_4H_{10}	2.175	-5082.13	-5097.05	-17.10	8605

Table 4. Ecal* (K) for 10 adsorbate molecules for P, PP, DPP models

Name	Formula	P	PP	PP	PP	PP	PP	DPP	DPP	DPP	DPP
Plate Separation (nm)			0.94	0.89	0.84	0.79	0.74	1.04	0.99	0.89	0.84
methane	CH_4	1749	1901	1978	2122	2352	2767	2012	2300	2923	2287
ethane	C_2H_6	2382	2754	2976	3329	3930	4544	3142	3501	4657	3859
chloromethane	CH_3Cl	2505	2857	3221	3414	3922	4507	3174	3628	4700	4356
chlorodifluoromethane	$CHClF_2$	2789	3248	3678	3972	4478	4587	3683	4410	5513	5255
1,1-difluoroethane	$C_2H_4F_2$	2585	3060	3346	3846	4491	5164	3510	4146	5401	5297
propane	C_3H_8	3363	3802	4054	4609	5368	6313	4315	5028	6672	5402
hexafluoroethane	C_2F_6	2782	3736	4324	5062	5469	4513	4286	5128	7074	7021
dichlorodifluoromethane	CCl_2F_2	2934	4322	4659	5155	5570	4035	4958	5932	7390	7038
1-chloro-1, 1-difluoroethane	$C_2H_3ClF_2$	2956	4066	4550	5160	5620	4783	5332	6034	7315	7490
butane	C_4H_{10}	4177	4814	5197	5957	7014	8071	5557	6588	8605	7522

In prior work, the gas phase adsorption of 118 organic molecules on graphite was studied by using MM2 molecular mechanics calculation to find molecule-surface binding energies, Ecal* [32]. Binding energy values used a three-layer model as in the plate model presented above. Experimental binding energies for graphite, E*(graphite), were available from gas-solid chromatography in the Henry's Law region over a range of temperature. When the 118 experimental E* values were correlated with the 118 MM2 calculated Ecal* values, it was found that a simple adjustment to MM2 results based on molecular structure improved the correlation and gave excellent results. This previous work showed that the linear relationship between E* and Ecal* was optimized by

$$E^* = \alpha \, Ecal^* \tag{37}$$

where multiplying Ecal* by a coefficient or equation represented by α better correlated to E*. Therefore the calculated binding energy should be modified as done in prior graphite study [32].

In the prior work, where the gas phase adsorption of 118 organic molecules on graphite was studied by using MM2 molecular mechanics, an equation for alpha in Eq. (37) was found. All non-hydrogen atoms of a molecule were separated into two categories: carbon sp^3 hybridized and all other atoms (carbon sp^2, carbon sp, oxygen, sulfur, nitrogen, fluorine, chlorine, etc.). Hydrogen atoms were excluded. The two categories were explained by the following equations

$$n_{total} = n_{C\text{-}sp^3} + n_{other} \tag{38}$$

$$f_{C\text{-}sp^3} = n_{C\text{-}sp^3} / n_{total} \tag{39}$$

$$f_{other} = n_{other} / n_{total} \tag{40}$$

$$1 = f_{C\text{-}sp^3} + f_{other} \tag{41}$$

where n_{total} is the total number of non-hydrogen atoms, $n_{C\text{-}sp^3}$ is the number of carbon sp^3 hybridized atoms, n_{other} is the number of all other nonhydrogen atoms in the molecule, $f_{C\text{-}sp^3}$ is the fraction of sp^3 carbon atoms in the molecule, and f_{other} is the fraction of other atoms in the molecule. The modified Ecal* is calculated by the following equation

$$Ecal^*_{modified} = (0.8180\ f_{C\text{-}sp^3} + 1.0221\ f_{other})\ Ecal^* \tag{42}$$

where 0.8180 and 1.0221 are best-fit coefficients derived from multi-linear regression calculations for the fraction of carbon sp^3 atoms and other non-hydrogen atoms [32].

From the prior result, notice that the factor alpha is given by

$$\alpha = (0.8180\ f_{C-sp^3} + 1.0221\ f_{other}) \tag{43}$$

where

$$Ecal^*_{modified} = \alpha\ Ecal^* \tag{44}$$

Eq. (42) was determined from an analysis of experimental data for the interaction energy of 118 organic molecules adsorbed on graphite E*, calculations of model molecule-surface interactions values Ecal*, and from comparisons of 118 pairs of data [32]. It was observed that the direct vdW interaction calculated from MM2 parameters with no modification agreed extremely well with experimental results if the molecules did not contain sp^3 C atoms. In these cases we expect only about a 2% modification by multiplying by 1.0221 to slightly increase the MM2 calculated interaction energy. However, the fraction of nonhydrogen sp^3 carbon atoms was found to cause calculated MM2 value to be an overestimate and had to be multiplied by 0.8180 to reduce the interaction energy result. When this corrective formula was applied in the prior work the E*(graphite) = 1.0000 $Ecal^*_{modified}$ (plate model) with $R^2 = 0.9647$ for 118 data points.

This rather significant result for organic molecules interacting with a graphite surface and using a plate model suggest that the same correction should be applied with work with more complex carbon surface models based on combinations of plates. That situation applies in the current work so we use Eqs. (42) and (43) to find modified values of Ecal*. The calculation for the surface DPP (D = 0.89nm) presented in Table 3 was repeated for each of the other nine surface models. For the 10 molecules and 10 surfaces, that generated a total of 100 different values of Ecal*. The 100 values of Ecal* were then converted to the 100 $Ecal^*_{modified}$ values shown in Table 5. Eq. (42) was used to convert to $Ecal^*_{modified}$. The 10 molecules on the various plate, parallel plate, and double parallel plate models are listed in Table 5. The MM2 calculated binding energy ($Ecal^*_{modified}$) values for the 10 molecules (n = 10) can now be compared to the experimental binding energy (E*).

Table 5. Experimental E* for 10 adsorbates and Ecal*$_{modified}$ (K) for 10 adsorbate molecules for each P, PP, DPP model

Name	**Formula**	**E* (K)**										
Model			**P**	**PP**	**PP**	**PP**	**PP**	**PP**	**DPP**	**DPP**	**DPP**	**DPP**
Plate Separation D (nm)				0.94	0.89	0.84	0.79	0.74	1.04	0.99	0.89	0.84
methane	CH_4	2287	1431	1555	1618	1736	1924	2264	1646	1881	2391	1870
ethane	C_2H_6	3859	1948	2253	2434	2723	3214	3717	2570	2864	3809	3157
chloromethane	CH_3Cl	4356	2305	2629	2963	3141	3609	4147	2920	3338	4324	4008
chlorodifluoromethane	$CHClF_2$	5255	2708	3154	3572	3857	4348	4454	3576	4282	5354	5103
1,1-difluoroethane	$C_2H_4F_2$	5297	2378	2815	3078	3539	4132	4751	3229	3815	4969	4874
propane	C_3H_8	5402	2751	3110	3316	3770	4391	5164	3530	4113	5458	4419
hexafluoroethane	C_2F_6	7021	2702	3628	4199	4916	5311	4383	4162	4980	6869	6817
dichlorodifluoromethane	CCl_2F_2	7038	2879	4241	4572	5058	5466	3960	4865	5821	7252	6906
1-chloro-1, 1-difluoroethane	$C_2H_3ClF_2$	7490	2780	3824	4279	4853	5285	4498	5015	5675	6879	7044
butane	C_4H_{10}	7522	3417	3938	4251	4873	5738	6602	4546	5389	7039	6153

Table 6. Ratio of $Ecal^*_{modified}/E^*$ for 10 molecules for models P, PP, and DPP

Name Model	Formula	P	PP	PP	PP	PP	PP	DPP	DPP	DPP	DPP
Plate Separation D (nm)			0.94	0.89	0.84	0.79	0.74	1.04	0.99	0.89	0.84
methane	CH_4	0.626	0.680	0.708	0.759	0.841	0.990	0.720	0.823	1.046	0.818
ethane	C_2H_6	0.505	0.584	0.631	0.706	0.833	0.963	0.666	0.742	0.987	0.818
chloromethane	CH_3Cl	0.529	0.603	0.680	0.721	0.828	0.952	0.670	0.766	0.993	0.920
chlorodifluoromethane	$CHClF_2$	0.515	0.600	0.680	0.734	0.827	0.848	0.681	0.815	1.019	0.971
1,1-difluoroethane	$C_2H_4F_2$	0.449	0.531	0.581	0.668	0.780	0.897	0.610	0.720	0.938	0.920
propane	C_3H_8	0.509	0.576	0.614	0.698	0.813	0.956	0.653	0.761	1.010	0.818
hexafluoroethane	C_2F_6	0.385	0.517	0.598	0.700	0.757	0.624	0.593	0.709	0.978	0.971
dichlorodifluoromethane	CCl_2F_2	0.409	0.603	0.650	0.719	0.777	0.563	0.691	0.827	1.030	0.981
1-chloro-1, 1-difluoroethane	$C_2H_3ClF_2$	0.371	0.511	0.571	0.648	0.706	0.601	0.669	0.758	0.918	0.940
butane	C_4H_{10}	0.454	0.524	0.565	0.648	0.763	0.878	0.604	0.716	0.936	0.818

Table 7. Average ratio of $Ecal*_{modified}/E*$ for P, PP, and DPP models for varied internuclei separation (D) and interior pore distance (d)

Model	D (nm)	d (nm)	Average ratio of $Ecal*_{modified}/E*$	Standard Deviation	Relative Standard Deviation (%)
P			0.475	0.077	16.3
PP	0.94	0.60	0.573	0.053	9.3
PP	0.89	0.55	0.628	0.050	8.0
PP	0.84	0.50	0.700	0.036	5.2
PP	0.79	0.45	0.792	0.044	5.5
PP	0.74	0.40	0.827	0.166	20.0
DPP	1.04	0.70	0.656	0.041	6.3
DPP	0.99	0.65	0.764	0.044	5.8
DPP	0.89	0.55	0.986	0.043	4.4
DPP	0.84	0.50	0.898	0.071	8.0

5. DISCUSSION

To compare the calculated binding energy with the experimental binding energy, $Ecal*_{modified}$ was divided by E* for each molecule for all 10 models. The ratio of $Ecal*_{modified}/E*$ was then calculated for each molecule with each model. Table 6 lists the ratio of $Ecal*_{modified}/E*$ for the 10 molecules for all P, PP, and DPP models. The optimal result would be a column with 1.00 for all values. This table supports that the DPP with a plate separation of 0.89 nm is the best model because the ratio of $Ecal*_{modified}/E*$ is closest to one for the 10 molecules. This ratio indicates that this computational work compares well to the experimental if the right surface model is selected.

Table 7 gives the average ratio of $Ecal*_{modified}/E*$ for each model, the standard deviation of the ratio for each model, and the standard deviation as a percent of the average or relative standard deviation (RSD). From Table 7, the results show that the plate model does not compare well because the average $Ecal*_{modified}/E* = 0.475$ with RSD = 16.3%. When the parallel plate model is used, the results are better but there is room for improvement because when D = 0.84 nm, the average $Ecal*_{modified}/E* = 0.700$ with RSD = 5.2%. The model that compares the best is the double parallel plate when D = 0.89 nm. The average $Ecal*_{modified}/E* = 0.986$ with RSD = 4.4%. Note that the RSD at 4.4% is lowest for the DPP (D = 0.89) pore model. Also for this model the average of the ratio at 0.986 is closest to one.

In a further analysis, 10 plots were made of E* versus $Ecal*_{modified}$ for the 10 adsorbate molecules. A total of 10 models were used including: P, PP (D = 0.94 nm), PP (D = 0.89 nm), PP (D = 0.84 nm), PP (D = 0.79 nm), PP (D = 0.74 nm), DPP (D = 1.04 nm), DPP (D = 0.99 nm), DPP (D = 0.89 nm), and DPP (D = 0.84 nm). Plots were made of E* = slope $Ecal*_{modified}$. For the P model the slope = 2.223 and R^2 = 0.784. So the E* values are more than twice the $Ecal*_{modified}$. In the PP models the slopes were 1.798, 1.631, 1.451, and 1.289 for plate separations of 0.94, 0.89, 0.84, and 0.79, respectively. The R^2 values were 0.931, 0.942, 0.972, and 0.967, respectively. So the best separation judging by the slope being closest to 1 is PP (D = 0.74 nm) but the best by R^2 being closest to 1 is PP (D = 0.79 nm).

In the DPP models the slopes were 1.543, 1.318, 1.025, and 1.091 for plate separations of 0.99, 0.94, 0.89, and 0.84, respectively. The R^2 values were 0.957, 0.961, 0.978, and 0.934, respectively. So the best separation judging by the slope being closest to 1 and the best by R^2 being closest to 1 is DPP (D = 0.89 nm). Of all 10 models examined, clearly the best results are for the DPP (D = 0.89 nm) that gives E* = 1.025 $Ecal*_{modified}$ with R^2 = 0.978.

Table 8. Results from plots of E* versus $Ecal*_{modified}$ for 10 gas molecules on P, PP, and DPP models for varied internuclei separation (D) and interior pore distance (d)

Model	D(nm)	d(nm)	slope	R^2
P			2.223	0.784
PP	0.94	0.60	1.798	0.931
PP	0.89	0.55	1.631	0.942
PP	0.84	0.50	1.451	0.972
PP	0.79	0.45	1.289	0.967
PP	0.74	0.40	1.256	0.485
DPP	1.04	0.70	1.543	0.957
DPP	0.99	0.65	1.318	0.961
DPP	0.89	0.55	1.025	0.978
DPP	0.84	0.50	1.091	0.934

In these E* versus $Ecal*_{modified}$ correlations for plate, parallel plate, and double parallel plate, the R^2 values show that the data points for the plate model are more spread out R^2 = 0.784 than most of the PP or all the DPP models. With the exception of one of the parallel plate models all the other PP R^2 are 0.931 or better. For the double parallel plate model, all the DPP R^2 are

0.934 or better. The very low R^2 indicating a poor data correlation for PP (D = 0.74) is due to the very small pore size. As the pore diameter becomes too small, the larger molecules in the set experience less attractive interaction or even repulsion within the pore.

Table 8 lists the slope and R^2 values for the 10 models for plots E* versus $Ecal*_{modified}$. If the experimental and calculated binding energies are similar then the slope and R^2 values will be close to one. From Table 8, the chart shows that E* and $Ecal*_{modified}$ do not compare well for the plate model because the slope is equal to 2.223 and R^2 is equal to 0.784. E* and $Ecal*_{modified}$ have the most similar values for the double parallel plate when D = 0.89 nm because the slope equals 1.025 and a value of 0.978 for R^2. The DPP (D = 0.89) is best of the 10 models examined. Figure 4 shows a plot of E* versus $Ecal*_{modified}$ for the best fitting model. This best fit model agrees with size exclusion data and N_2 pore analysis giving the Horvath-Kawazoe differential pore volume plot maximum at 0.55nm [1].

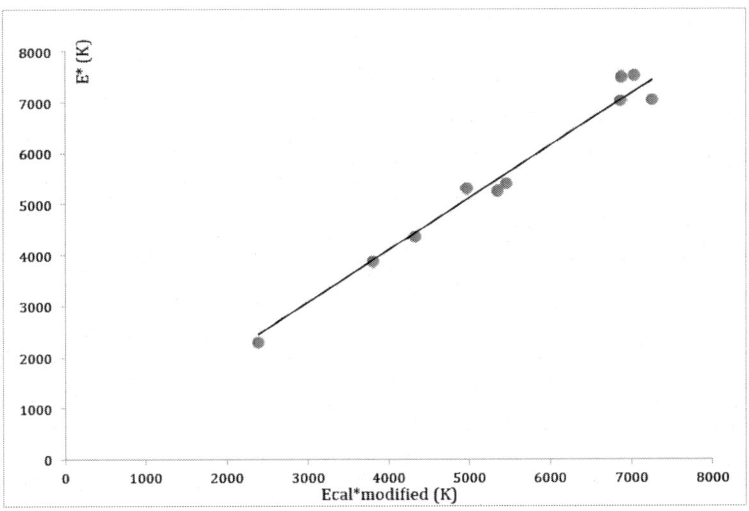

Figure 4. Plot of molecule–surface binding energy E*(K) versus $Ecal*_{modified}$ (K) for Double Parallel Plate model with internuclei separation of D = 0.89 nn. Linear regression of E* = 1.025 $Ecal*_{modified}$ with R^2 = 0.978.

Thermal programmed desorption (TPD) has been used to determine monolayer coverage binding energies for polyaromatic hydrocarbons (PAH) [33] on a sample of highly oriented pyrolytic graphite (HOPG) [34]. TPD data provide an interesting comparison to the GSC data since it is necessary to include adsorbate-adsorbate interactions as well as adsorbate-surface interactions. Thermal desorption spectra at monolayer coverage were converted to binding energy values, E*, in this work by Zacharia et al. [33] and Ulbricht et al. [34]. MM2 mechanics based calculations, led to an improved correlation between E* and Ecal* [16] relative to that previously reported for the TPD data [34]. The graphitic surface was represented by three graphene layers with 702 benzene rings and 1,510 carbon atoms per layer. Their TPD E* values spanned a range of from 1918K to 25257K (0.17 to 2.18eV).

A complete understanding of porous carbons, graphene surfaces or carbon nanotubes would require a quantum mechanical description and inclusion of dispersion forces and long range electron correlation [35]. Ramos-Berullas et al. used pyridine and adsorption energies were obtained using different DFT functionals including implicitly dispersion-corrected functionals, namely the M06-2X dispersion corrections. They studied the effect of the surface curvature on the binding strength and interaction energy [35]. While these more careful theoretical considerations are significant, molecular mechanics continues to have a role in making practical estimates of molecule-carbon surface binding energy as in our current work.

MM2 parameters for atom-carbon interactions prove to be effective in simple calculations of molecule-surface binding energies. A flat graphitic surface can be well represented by parallel graphene structures with a number of interconnected 6 member aromatic (benzene) rings. Other surface structures with interesting geometries can be modeled. Molecular mechanics offers the speed and simplicity to use hundreds or thousands of atoms in the surface structure and place multiple molecules on a surface. The energy differences between near and far molecule-surface separation can be provide useful estimates of adsorption binding energies. The molecular mechanics method will not elucidate weak dispersive interaction details or electronic details, but does provide some useful applications. For example, simpler calculations of molecule-surface binding energies continue to be of interest to correlate with and predict experimental binding energies.

CONCLUSION

Gas-solid chromatography experiments were performed to measure second gas-solid virial coefficients over a range of temperatures for a set of 10 organic molecules interacting with Carbosieve S-III carbon powder. The experimental molecule-surface binding energies (E*) were found from van't

Hoff plots of ln(B_{2s}) versus 1/T for each molecule. The slope of each plot is equal to the experimental binding energy. Our interest was to be able to correlate these E* values with calculated molecule-surface biding energy values modified by a simple correction to the MM2 molecular mechanics results. The correction primarily adjusts for an underestimate of the vdW interaction between the adsorbate molecule sp^3 carbon atoms and the surface carbons.

The modified calculated MM2 interaction ($Ecal^*_{modified}$) provided an excellent agreement with the experimental E* values. For the best surface model the linear regression gave $E^* = 1.025\ Ecal^*_{modified}$ with a value of 0.978 for R^2. This best surface model was a double parallel plate model with a square pore of internuclei separation of D = 0.89 nm and d = 0.55 nm for the internal separation. A total of 10 models were examined. The models represented flat surface, two-sided pores and four-sided pores or plate (P), parallel plate (PP), and double parallel plate (DPP) models, respectively. In order to find the best model, each model was tested by comparison to experimental data.

This approach allows predictions of other molecule-surface binding energies for Carbosieve S-III. In addition to the benefit of using molecular mechanics to be able to predict experimental values of interest, we are also able to extract some basic information about the porosity of the powder studied. While details are lost in a process that generates an idealized model surface, it is easy to distinguish a porous from a nonporous surface. The plate model gives an average ratio for the 10 molecules of $0.475 = E^*/Ecal^*_{modified}$ with a 16.3% relative standard deviation. This result indicates that the nonporous model does not represent the full molecule-surface interaction. Only a model surface with small pores can reproduce the full E* interaction for the set of molecules. The best DPP model was a pore width size of d = 0.55 nm for Carbosieve S-III that agreed with prior experimental pore distribution data. An average ratio for the 10 molecules using this best DDP model gave $0.986 = E^*/Ecal^*_{modified}$ with a relative standard deviation of 4.4%. This indicates an appropriate and consistent prediction of E* for the 10 molecules studied.

While our calculations cannot capture the details of pore distribution data, our research here and previously showed that MM2 computational average molecule-surface binding energies can be generated that successfully predict experimental values. Once correlations of $Ecal^*_{modified}$ with E* are determined, predictions of binding energies for additional molecules interacting with the same adsorbent can be made. This approach is advantageous where no experimental data are available. For compounds without sp^3 carbons (aromatic or substituted aromatic) the Ecal* directly obtained from MM2 calculations with no further modifications are reliable estimates of the molecule-surface interaction. The binding energies in this work (the E*, Ecal*, and $Ecal*_{modified}$) are in the limit of low coverage where adsorbate molecules are interacting with the surface and not each other.

Newer carbon and other nanoporous materials have found use in varied applications including molecular isolation, separation, storage, sensor effects, and nanoscale reactions. While electronic details are revealed only in more sophisticated quantum calculations, the practical and quicker applications of molecular mechanics make this a technique that can still provide useful insights into roles of noncovalent molecular interactions with surfaces and predictions of practical values.

Future work for this research could include calculations for other adsorbate-adsorbent systems including carbon nanotubes and graphene layered materials. An excellent review providing details of inverse gas chromatography (IGC) and interpreting chromatographic enthalpies and entropies for varied adsorbates on carbon adsorbent materials has been presented [36]. Acquiring experimental B_{2s}, and subsequent determinations of E* and then correlating experimental and calculated molecule-surface binding energies may lead to better understanding of adsorbate-adsorbent based applications. Utilizing and predicting the role of of E*, B_{2s}, and related thermodynamic properties with molecular mechanics models may lead to better practical applications of molecule separation and storage.

ACKNOWLEDGMENT

We acknowledge the support provided by the University of Tennessee at Chattanooga Grote Chemistry Fund and the Furman University Summer Undergraduate Research Program.

REFERENCES

[1] Rybolt, TR; Ziegler, KA; Thomas, HE; Boyd, JL; Ridgeway, ME. *Journal of Colloid and Interface Science*., 2006, 296, 41-50.

[2] Lu, GQ; Zhao, XS. In *Nanoporous Materials: Science and Engineering*; Imperial College Press, London, 2004, Vol. 4, 1-13.
[3] Jankowska, H; Swiatkowski, A; Choma, J. *Active Carbon*; Ellis Horwood: New York, NY, 1991,76-103.
[4] Zhao, Q; Qian, F; Zhu, L; Chen, Z; Wang, S. *Zhongguo Huanjing Jiance*., 2014, 30(3), 107-110.
[5] Yunes, S; Wommack, P; Still, M; Kenvin, J; Exley, J. *Appl. Catal. A*., 2014, 474, 250-256.
[6] Jezewska, A. *Podstawy i Metody Oceny Srodowiska Pracy*., 2013, 29(2), 155-170.
[7] Tassi, F; Capecchiacci, F; Buccianti, A; Vaselli, O. *Applied Geochemistry*., 2012, 27(1), 115-123.
[8] Seo, YS; Lee, JK. *Hwahak Konghak*., 2010, 48(2), 244-252.
[9] Rybolt, TR; Thomas, HE. In *Interfacial Phenomena in Chromatography*; Pefferkorn, E; Ed; Dekkar: New York, NY, 1999, 1-40.
[10] Allinger, NL. *J. Am. Chem. Soc*., 1977, *99*, 8127-8134.
[11] Allinger, NL; Yuh, YH; Lii, J. *J. Am. Chem. Soc*., 1989, *111*, 8551-8566.
[12] Lii, J; Allinger, NL. *J. Am. Chem. Soc*., 1989, *111*, 8576-8582.
[13] Jensen, F. *Introduction to Computational Chemistry*; John Wiley & Sons: New York, NY, 1999.
[14] Rappe, AK; Casewit, CJ. *Molecular Mechanics across Chemistry*; University Science Books: Sausalito, CA, 1997.
[15] Rybolt, TR; Hansel, RA. *J. Colloid Interface Sci*., 2006, *300*, 805-808.
[16] Rybolt, TR; Wells, CE; Sisson, CR; Black, CB; K. A. *J. Colloid Interface Sci*., 2007, 314, 434-445.
[17] Rybolt, TR; Wells, CE; Thomas HE; Goodwin, CM; Blakely, JL; Turner, JD. *J. Colloid Interface Sci*., 14, 325, 282-286.
[18] Rybolt, TR; Wells, CE. In *Graphene and Graphitic Materials*; Chan, H. E; Ed; Nova Science Publishers: New York, NY, 2009, 95-112.
[19] Burress, JW; Gadipelli, S; Ford, J; Simmons, JM; Zhou, WT. Yildirim. *Angew. Chem. Int. Ed*., 2010, 49, 8902-8904.
[20] Bogdan, K; Firlej, L; Mohammadhosseini, A; Boulet, P; Beckner, M; Romanos, J; Pfeifer P. *J. Am. Chem. Soc*., 2012, 134, 15130-15137.
[21] Cabria, I; Lopez, MJ; Alonso, JA. *Phys. Rev. B*., 2008, 78, 075415.
[22] Jiao, Y; Du, A; Hankel, M; Smith, SC. *Phys. Chem. Chem. Phys*., 2013, 15, 4832-4843.
[23] Jagiello, J; Olivier, JP. *Carbon*., 2013, 55, 70-80.

[24] Wongkoblap, A; Do, DD. *Mol. Simulation.*, 2006, 32, 539-549.
[25] Bandosz, TJ; Biggs, MJ; Gubbins, KE; Hattori, Y; Iiyama, T; Kaneko, K; Pikunic, J; Thomson, KT. *Chemistry and Physics of Carbon*, Marcel Dekker: New York, NY, 2003, 42-61.
[26] Goodman, JM. *Chemical Applications of Molecular Modelling*; The Royal Society of Chemistry: Cambridge, 1998, 1-120.
[27] Betz, WR. Maroldo, SG. Wachob, GD. Firth, MC. Am. Ind. Hyg. Assoc. J., 1989, 50, 181-187.
[28] Lucena, S; Goncalves, D; Mileo, P; Cavalcante Jr., C. *Adsorption.*, 2012, 18, 113-119.
[29] Rybolt, TR; Logan, DL; Milburn, M; Thomas, HE; Waters, AB. *J. Colloid Interface Sci.*, 1999, 220, 148-156.
[30] Olson, DR; Rybolt, TR; Bodine, DS; Wall, MD. *J. Colloid Interface Sci.*, 1993, 159, 205-209.
[31] Rybolt, TR; Bivona, KT; Thomas, HE; O'Dell, CM. *J. Colloid Interface Sci.*, 2009, 338, 287-292.
[32] Son, JH; Rybolt, TR. *Graphene*. 2013, 2, 18-34.
[33] Zacharia, R; Ulbricht, H; Hertel, T. *Phys. Rev. B.*, 2004, 69, 155406/1-155406/7.
[34] Ulbricht, H; Zacharia, R; Cindir, N; Hertel, T. *Carbon.*, 2006, 44, 2931-2942.
[35] Ramos-Berdullas, N; Pérez-Juste, I; Alsenoy, CV; Mandado, M. *Phys. Chem. Chem. Phys.*, 2015, 17, 575-587.
[36] Grajek, HJ; Chromat A., 2007, 1145, 1-50.

INDEX

B

C

D

E

F

G

H

I

K

L

M

N

O

P

Q

R

S

T

V

W

Y

Z